CHEMISTRY RESEARCH AND APPLICATIONS

OCHRATOXIN A AND AFLATOXIN B1

NEW RESEARCH

Chemistry Research and Applications

Additional books and e-books in this series can be found on Nova's website under the Series tab.

CHEMISTRY RESEARCH AND APPLICATIONS

OCHRATOXIN A AND AFLATOXIN B1

NEW RESEARCH

REUBEN HESS

EDITOR

nova

science publishers

New York

Library of Congress Cataloging-in-Publication Data

ISBN: 978-1-53617-416-8
Library of Congress Control Number: 2020931181

Published by Nova Science Publishers, Inc. † New York

Contents

Preface vii

Chapter 1 Ochratoxin A (OTA) Occurence in Meat and Dairy Products: Prevention and Remediation Strategies **1**
D. Pattono and F. Gai

Chapter 2 Ochratoxin A Contamination of Traditional Dry-Cured Meat Products **31**
Jelka Pleadin, Manuela Zadravec, Tina Lešić, Ana Vulić, Nada Vahčić, Nina Kudumija, Jadranka Frece and Ksenija Markov

Chapter 3 Co-Occurrence of Ochratoxin A and Citrinin and Identification of Fungi That Produce the Respective Mycotoxins in Mold-Contaminated Rice, Corn and Groundnut Samples **67**
L. Gayathri, B. Karthiyayini, K. Ruckmani, D. Dhanasekaran and M. A. Akbarsha

Chapter 4 Presence of Aflatoxin B1 in Beer: Contamination during Processing and Methods of Analysis **101**
Antonio Ruiz-Medina

Index **127**

PREFACE

Ochratoxin A and Aflatoxin B1: New Research first presents the latest knowledge regarding the occurrence of ochratoxin A in dry-cured meat and dairy products, as well as relevant food safety issues. Different remediation strategies such as heating, ripening, drying, storage and the use of microbial cultures, which have been adopted for the prevention or decontamination of ochratoxin A, are critically reviewed.

Next, the authors assess the occurrence of ochratoxin A in different types of dry-fermented sausages and hams produced by various households in Croatia, proposing that in order to avoid contamination, the production of household-based meat and meat products should run under standardized and well-controlled conditions.

In the subsequent study, a total of sixty mold-contaminated rice, corn and groundnut samples were collected from one of the Cauvery delta regions of Tamil Nadu, India, and analyzed to determine the co-occurrence of ochratoxin A and citrinin, and the respective mycotoxigenic fungal strains that produce these toxins.

Besides the known adverse effects of alcohol, beer could also be the source of several mycotoxins such as aflatoxin B1. Aflatoxin B1 is the most toxic among the identified aflatoxins and classified into group I as a human carcinogen by the International Agency for Research on Cancers. The stability of aflatoxin B1 during the brewing process, as well as the most important detection techniques and sample treatment used for the determination of this compound are evaluated.

Chapter 1 - Ochratoxin A (OTA) is a significant mycotoxin produced by *Aspergillus* and *Penicillium* species that exhibits a toxic or potentially carcinogenic activity in animals and humans. OTA is known to widely contaminate feeds and food, including food of animal origin, such as milk, cheese and meat, thereby raising food safety issues. However, the European Commission has not yet established any maximum permitted limits of OTA in meat or other animal products, although some countries have enforced maximum levels of OTA concentrations or have developed guidelines in which maximum OTA levels are recommended. This chapter, after a brief introduction, is aimed at informing the readers about the latest knowledge about the occurrence and the food safety issues of OTA in dry-cured meat and dairy products. Moreover, the different remediation strategies, such as heating, ripening, drying, storage and the use of microbial cultures, which are adopted for the prevention or decontamination of OTA in the above mentioned food products, are critically reviewed.

Chapter 2 - Dry-cured meat products produced using traditional technological processes represent food brands best recognised worldwide. During ripening, the surface of these products becomes overgrown with moulds whose spores mostly come from the environment in which ripening chambers are placed. The intensity of the overgrowth is enhanced by ripening longevity and traditional production environment in which, usually, no microbiological filters and no pneumatic barriers are used, so that the air temperature and relative humidity are virtually uncontrollable. Surface moulds are mainly of the *Penicillium* and *Aspergillus* genera, some of those having beneficial effects on product quality, and some causing mycotoxin contamination of the final products. The presence of the mycotoxin ochratoxin A (OTA) in dry-cured meat products can be mould-generated or come as a consequence of an indirect contamination (*carryover* effect) in case the meat-providing animals had been exposed to contaminated diet during farming. Another path of contamination are contaminated ingredients (e.g., spices) used in dry-cured meat products' production. This study investigated into the occurrence of OTA in different types of dry-fermented sausages and hams (n = 135) produced by a large number of households situated in different Croatian regions and sampled from the markets. Surface moulds were isolated and identified using traditional mycological and molecular (polymerase chain reaction, PCR) methods. The presence of OTA was first determined using a validated immunoassay method (ELISA); in samples with OTA concentrations higher than the limit of detection, the mycotoxin presence was confirmed using liquid chromatography tandem mass spectrometry (LC-MS/MS). OTA presence was determined in 11% of samples in the maximal concentration of 2.54 μg/kg in a ham sample. From the products' surfaces, a total of 404 *Penicillium*, 102 *Aspergillus* and 15 *Mucor* isolates were retrieved, out of which *Aspergillus niger, Aspergillus ochraceus* and *Penicillium verrucosum,* known as OTA- producers. The results showed random OTA contamination of traditional dry-cured meat products, indicating that, in order to avoid such a contamination, household-based meat & meat products' production should run under standardized and well-controlled conditions.

Chapter 3 - OchratoxinA (OTA) and citrinin (CTN) are the most commonly co-occurring mycotoxins in a wide variety of food and feed commodities. As a combination these mycotoxins are reported to cause endemic nephropathy, hepatotoxicity and pulmonary toxicity. Hence, it is important to check the quality of food commodities that are destined for human and/or animal consumption, with special reference to occurrence/co-occurrence of mycotoxins. Therefore, in this study a total of sixty mold-contaminated rice, corn and groundnut samples were collected from one of the Cauvery delta regions of Tamil Nadu, India, and analyzed to find the co-occurrence of OTA and CTN, and the respective mycotoxigenic fungal strains that produce these toxins. Co-occurrence of OTA and CTN was qualitatively and quantitatively confirmed by thin layer chromatography (TLC) and High Pressure Liquid Chromatography (HPLC) methods. OTA and CTN co-occurred in 35%, 25% and 15% of the contaminated rice, corn and groundnut samples, respectively. From the mold-contaminated rice, corn and groundnut samples, a total of 60 morphologically distinct strains of *Aspergillus* sp and *Penicillium* sp were isolated and pre-screened for the production of mycotoxin(s) using coconut cream agar (CCA) plating method. Isolates of fungi that fluoresced highly were cultured in yeast extract sucrose (YES) broth and the production of OTA and CTN was analyzed using HPLC. OTA- and CTN-positive strains were characterized and identified by ITS gene sequence analysis as *Aspergillus tubingensis, A. flavus, A. niger,* and *A. oryzae*. Co-occurrence of OTA & CTN and identification of mycotoxigenic fungi that produce the respective mycotoxins present in rice, corn and groundnut indicate the potential health risk of these mycotoxins either singly or in combination to humans as well as animals.

Chapter 4 - Besides the known adverse effects of alcohol, beer could also be the source of several mycotoxins such as aflatoxin B1 (AFB1), the most toxic one among the identified aflatoxins and classified into group I as a human carcinogen by the International Agency for Research on Cancers. For this reason, the contamination of cereals and therefore beer is controlled worldwide by legal limits to ensure public health safety.

The appearance of AFB1 in beer is the result of contamination from malted grain or food additives. The fate and stability of AFB1 during brewing process (malting, mashing, fermentation, maturation, etc.) as well as the most important detection techniques and sample treatment used for the determination of this compound will be evaluated in this chapter.

In: Ochratoxin A and Aflatoxin B1
Editor: Reuben Hess
ISBN: 978-1-53617-416-8

Chapter 1

OCHRATOXIN A (OTA) OCCURENCE IN MEAT AND DAIRY PRODUCTS: PREVENTION AND REMEDIATION STRATEGIES

D. Pattono[1,*] ***and F. Gai***[2]

[1]Department of Veterinary Sciences,
University of Turin, Grugliasco, Italy
[2]Institute of Sciences of Food Production,
Italian National Research Council, Grugliasco, Italy

ABSTRACT

Ochratoxin A (OTA) is a significant mycotoxin produced by *Aspergillus* and *Penicillium* species that exhibits a toxic or potentially carcinogenic activity in animals and humans. OTA is known to widely contaminate feeds and food, including food of animal origin, such as milk, cheese and meat, thereby raising food safety issues. However, the

* Corresponding Author's Email: daniele.pattono@unito.it.

European Commission has not yet established any maximum permitted limits of OTA in meat or other animal products, although some countries have enforced maximum levels of OTA concentrations or have developed guidelines in which maximum OTA levels are recommended. This chapter, after a brief introduction, is aimed at informing the readers about the latest knowledge about the occurrence and the food safety issues of OTA in dry-cured meat and dairy products. Moreover, the different remediation strategies, such as heating, ripening, drying, storage and the use of microbial cultures, which are adopted for the prevention or decontamination of OTA in the above mentioned food products, are critically reviewed.

Keywords: ochratoxin family, ochratoxin A (OTA), food quality, food spoilage, food safety, meat products, dairy products, control policy

Introduction

Ochratoxin A (OTA) is a mycotoxin that is produced by some toxigenic mold species of the *Penicillium* and *Aspergillus* genera, and this metabolite is considered genotoxic due to the formation of OTA-DNA adducts (Ostry et al., 2016), but it has also shown nephrotoxic, hepatotoxic and immunotoxic properties (Schmidt-Heydt et al., 2011). In addition, this mycotoxin has been classified, by the International Agency for Research on Cancer, as Group 2B carcinogenic (possibly carcinogenic to humans) (Ostry et al., 2016).

OTA contamination of food commodities (i.e., cereals and cereal products, pulses, coffee, nuts and spices) has been reported all over the world (EFSA 2006). Furthermore, OTA can also contaminate infant formulas, baby foods, dried meats, blood sausages, meat and milk, and some molds, such as *Penicillium nordicum*, *Penicillium verrucosum* or *Aspergillus westerdijkiae,* are in particular the most frequently detected in dry-cured meat products because they are able to successfully adapt to the characteristic NaCl-rich environment (Alapont et al., 2014; Rodríguez et al., 2014; Vipotnik et al., 2017).

Although toxin production may occur over a wide temperature range, the optimal conditions for ochratoxin production fall within a temperature range of between 20 and 25 °C and a crop moisture content of at least 16% (Völkel et al., 2011).

In the same way as for all other mycotoxins, the contamination of agricultural products with OTA is difficult to predict, since it depends on a complex interaction of factors, such as moisture, temperature, storage conditions, product type, storage time and distribution process (Khalesi & Khatib, 2011).

In the following sections, we examine the contamination routes for milk and dairy products as well as for dry-cured meat products. Moreover, different remediation strategies, such as heating, ripening, drying, storage and the use of microbial cultures, which are adopted for the prevention or decontamination of OTA in the above mentioned food products, are critically reviewed.

Milk and Dairy Products

The contamination routes can be classified as either direct or indirect contamination. Direct contamination refers to a contamination that arises directly from mold growing in milk or dairy products, while indirect contamination indicates when a contaminated feed is ingested by animals (Hymery et al., 2014).

In the case of milk and dairy products, the role of the two different contamination routes does not have the same weight.

Milk, like other foods of animal origin, has generally rarely been indicated as being contaminated by OTA (Flores-Flores et al., 2015; Fink-Gremmels and Van der Merwe, 2019; Carballo et al., 2019), even though several OTA producing species have been isolated from raw milk (Younis et al., 2016). Younis et al. (2016), found mold incidence rates of about one third in samples of cow's milk (33.5%) and the most prevalent species was *A. niger* although other species of *Aspergillus* such as *A. flavus*, *A. terreus*, *A. versicolor* and *A. oryzae* have been found. It should be pointed out that contamination by the producing mold may be of great importance for dairy products, but be much less important for the milk itself as an edible product. The contamination of raw milk can take place during different phases of the production chain, such as the milking phase, transportation and storage (Younis et al., 2016). Despite the presence of producing molds, OTA has rarely been found in milk. Several studies have been conducted in industrialized and developing countries and the milk situation seems to be quite homogeneous. Cow's milk has been found positive to OTA at different percentages, as reported by Turkoglu and Keyvan (2019) : 0% in Germany, 14% in Sweden and 15% in Norway. One study conducted on cow's milk in Spain reported no contamination (Flores-Flores and Gonzáles-Peñas, 2018). China also reported a contamination percentage close to 0% (one sample out of 125), but in that study the author stressed that a multiple contamination was frequently detected in several samples (Mao et al., 2018). The same results were found in Egypt by Kamal et al. (2019), who sampled the raw milk of different species (cows, buffalos, sheep and goats) and no sample was found to be positive. However, their results were in contrast with the results of another study in which the percentage of contamination in cow's milk rose to 70%, but fewer samples were considered in the former study (1/3). Moreover, the detection method (ELISA *vs* Ultra Performance Liquid Chromatography – UPLC) could have played an important role (Younis et al., 2016). Higher positive percentages have also been detected in Turkey, where cow's milk has been found positive in 100% of the considered samples (Turkoglu and Keyan, 2019).

Breast milk can also be considered a source of OTA, as highlighted by several authors (Hof, 2016; Sengling Cebin Coppa et al., 2019). Cereals have been identified as the principal source of such a contamination.

Research has recently focused on donkey milk as an alleged alternative to breast milk, because of its composition, and for cases of intolerance to cow's milk (Conte and Panebianco, 2019). Considering the lack of detoxification protozoa (poligastric *vs* monogastric) and the easier adsorption in an acid stomach environment, Gross et al. (2019) analyzed donkey milk and. again in this case, no sample was found positive. The feeding habits of donkeys (grass, hay and minerals) can explain the extremely low percentage. However, different positive percentage can be hypothesized in the case of professional donkey breeding farms where large amount of cereals and cereal-based pellets, a relevant source for mycotoxin intake, are consumed (Gross et al., 2019). Because there have been fewer studies on this topic for donkeys than for other milk producing species, more research is needed to establish a reliable Hazard Analysis (Conte and Panebianco, 2019; Gross et al., 2019).

In most of the studies, low quantities have been found that ranged from below the LOD, depending on the used detection method, to 0.11 μg/l (Becker-Algeri et al., 2016; Kamal et al., 2019; Turkoglu and Keyan, 2019). This is not surprising because, due to the basic characteristics of OTA, it is slowly absorbed in the rumen, and the main detoxification system for poligastric animals is the rumen protozoal flora which is responsible for the degradation of OTA into a less toxic compound, that is, OTα (Becker-Algeri et al., 2016; Tao et al., 2018; Taheur et al., 2019). Nevertheless, even at low quantities, the presence of OTA should be considered and analyzed because milk is, for a specific range of consumers, the only food (Flores-Flores et al., 2015; Kamal et al., 2019) and also because even low contamination levels can contribute to the "tolerable week intake" (TWI) for consumers with specific habits: *eg* heavy coffee drinkers (Becker-Algeri et al., 2016; Hof, 2016).

In addition to milk, milk-based products, and in particular milk-based formulas, can also be a source of ingestion and be a source of criticality for a certain range of consumers. These formulas are important because they can be another source of exposure for infants to OTA in addition to cereal-based baby formulas. Albeit in different years, 24% (mean concentration of 0.50 ng/g) and 19% (concentrations ranging from 0.017 to 0.184 ng/g – mean concentration of 0.103 ng/g) of milk-based infant formulas have been found positive throughout the world (Capozzo et al., 2017). A similar percentage (20%) was calculated in Portugal (mean concentration of 0.135 ng/g). A percentage of 72% was calculated in Italy, with concentrations ranging from 0.04 to 0.69 ng/g (Capozzo et al., 2017; Elaridi et al., 2019). In the USA, milk-based baby formulas have always been found negative (Capozzo et al., 2017). In Lebanon, 91% of the considered baby formulas have recently been found positive to OTA, with a mean contamination of 0.37 ± 0.10 μg/kg. Fortunately, considering the Tolerable Daily Intake (TDI), the total amount has always been under the provisional daily intake calculated by EFSA and JEFCA (Elaridi et al., 2019).

Cheese is also considered a possible source of contamination for humans, even though it does not seem to be of great importance for this matrix considering the general population intake (Hymery et al., 2014). It can be assumed that this evidence is also the consequence of a lack of studies in literature on this mycotoxin in comparison to other mycotoxins, such as Aflatoxins. However, OTA producing molds have been found in several different cheeses (Decontardi et al., 2017; Anelli et al., 2019; Ramos-Pereira et al., 2019).

Indirect contamination seems to play a minor role for OTA than direct contamination, through the direct production of toxins in the matrix or on the rind. This statement is clearly supported by the fact that the distribution of OTA in cheese forms is generally not homogeneous and the rind is frequently the only or the more contaminated part (Dall'Asta et al., 2008; Anelli et al., 2019).

The presence and the role of OTA in cheese were reviewed by Hymery et al. (2014). Mycotoxin producing species in cheese can generally play a competing role with other microorganisms. They act as inhibitors of the growth and reproduction of competitor flora. This role could explain why toxin producing species or strains in cultural media and under laboratory conditions (no exposure to competitive stressing condition) frequently lose the ability to produce toxins (Coton et al., 2019). Another reason may be linked to the fitness of the species or the strain. Toxin production could be the expression of an adaptation of the organism to particular conditions of the matrix. For example, the production of OTA by *P. nordicum* or *P. verrucosum* has been hypothesized to be a metabolic way of ensuring a constant excretion of chlorine from the cell. Both species in fact produce higher quantities of OTA in salt-rich habitats (Camardo Leggieri et al., 2017; Coton et al., 2019). OTA production can also correspond to a waste elimination process (Bagher Hashemi and Gholamhosseinpour, 2019; Chiocchetti et al., 2019).

Several studies about the conditions necessary for OTA producing species to grow and to procure toxins in fresh cheese and under ripening conditions are present in literature. Camardo Leggieri et al. (2017) examined the growth and toxin producing conditions of *Penicillium nordicum* and *P. verrucosum* at different temperatures and A_w. *P. verrucosum* was found to be able to produce OTA at temperatures ranging from 15°C to 30°C at A_w 0.99, with a maximum rate at 25°C, while the temperatures for *P. nordicum* ranged from 10°C to 30°C at A_w 0.99 (maximum rate at 20°C). In addition, *P. nordicum* was also able to produce OTA at Aw 0.96. This is important for some long ripening types of PDO cheese, such as Grana Padano and Parmigiano Reggiano, or for some dry-cured meat products (Hymery et al., 2014; Sánchez-Montero et al., 2019a).

Another recent study has demonstrated the ability of *P. verrucosum* to produce OTA in a semi-hard PDO cheese called Comté (Coton et al., 2019).

The production of OTA began after 28 days at 8°C and reached a maximum quantity (3500 ng/g) at 35 days in the deeper layers of the cheese (10 mm). The authors already detected OTA at 7 days at 20°C, but no variation in quantity or depth was noted when lower temperatures were considered.

Unfortunately, despite all this evidence, not much literature is available about the detection of OTA in commercially made cheese.

Dall'Asta et al. (2008) reported OTA in several Blue Cheeses for the first time. They found 30 out of 92 (32.6%) samples of Gorgonzola and Roquefort Cheese positive to OTA at levels ranging from 0.1 μg/kg up to 03.0 μg/kg (mean of 0.69 μg/Kg – median (0.35 μg/kg). OTA was also quantified in semi hard artisanal cheese from North Italy at different levels. A total of 6 out of 32 samples (corresponding to 32%) were positive, and the values of the samples of the inner parts varied from 18.4 μg/kg to 146.0 μg/kg, while those of the rind samples varied from 1.0 μg/kg to 262.2 μg/kg (Becker-Algeri et al., 2016). Anelli et al., (2019) have recently reported the same results for Cave cheese, a traditional artisanal ripened cheese from South Italy. The cheese forms were divided into three subsamples and analyzed considering three different depth: the rind (till a depth of 2 cm), the middle and the central part. They found 36% of the rind samples were positive to OTA, but none of the deeper parts were positive.

In another study, Younis et al., (2016) found very high positive rates of soft (Kariesh) and hard (Roomy) cheese in Egypt at 80% and 90%, respectively. The level ranged between 0.045 ppm and 0.2 ppm (mean of 0.088 ± 0.024) in the soft cheese and between 0.01 ppm and 0.021 ppm (mean 0.015 ± 0.001) in the Roomy cheese.

Dry-Cured Meat Products

Dry-cured meats products (i.e., dried fermented salami and other products) are consumed in various regions of the world and consumers are increasingly demanding better quality and safer products because they are prone to contamination by certain toxigenic mold species (Ferrara et al., 2015; Lippolis et al., 2016).

As far as the cause of OTA contamination of dry-cured meats and dry-fermented products is concerned, several studies have shown that this type of contamination can occur at different points of the production chain, from the field (animals contaminated through the feeds) to the production or storage of the final product, as recently reviewed by Montanha et al. (2018).

OTA can reach the meat and offal of farm animals through OTA-contaminated feeds and, consequently, the final meat products (Perši et al. 2014; Pleadin et al. 2015). The addition of contaminated spices or the addition of yeast to these foods, to improve the flavor of the products (Perrone et al., 2019), as well the ingestion of contaminated products are in general the main routes of contamination.

Contamination depends to a great extent on the characteristics and the environmental conditions of the manufacturing plants, particularly with reference to temperature, relative humidity, and the composition of the environmental mycoflora (Iacumin et al., 2009; Rodríguez et al., 2015a).

Although fungal development contributes to the desired sensory characteristics, some molds, such as *Penicillium nordicum*, are able to produce OTA in meat products and dry-cured meat products are usually contaminated with molds during ripening due to improper drying or rehydration during storage (Asefa et al., 2011; Markov et al., 2013).

Mycotoxins were detected in 64% of 90 samples of Croatian meat products, and among the detected mycotoxins, OTA was the predominant contaminant (Markov et al., 2013), while Iqbal et al. (2014) found a maximum OTA concentrations of 7.83 μg Kg^{-1} in commercial salami samples. In another study carried out in Italy, on a total of 110 samples of different types of cured hams, an OTA concentration of 0.53 μg Kg^{-1} was found on the surface of the product in 84 samples, while a concentrations of less than 0.1 μg.Kg^{-1} was encountered in the innermost layers of the product in 32 samples (Dall'Asta et al., 2010). OTA was detected in blood sausages and liver-type sausages produced in Germany (Gareis and Scheuer, 2000). OTA was also found in salami and in dry-cured Iberian ham in Spain (Rodríguez et al., 2015b), but in concentrations below the LOD and LOQ (Bernáldez et al., 2013).

Although the presence of mycotoxins in dry-cured meats has been reported in several regions of the world, the presence of these contaminants is not regulated in most countries. Therefore, it is important to put in place methods to identify and reduce the contamination of dry-cured meats, in order to minimize the consumption of and deleterious effects caused by mycotoxins.

CONTROL AND DETOXIFICATION METHODS FOR MEAT AND DAIRY PRODUCTS

The presence of mycotoxins in feeds and food is a fact of great concern due to their effects on human and animal health and to the resulting economic export losses (Milicevic et al., 2015). For these reasons, their control has always been a goal for producers, Health Services and the scientific community (Capozzo et al., 2017). At the moment, only European Regulation 2015/786 deals with decontamination of feed products from mycotoxins, and Karlowsky et al. (2016) also proposed this regulation as a guide for food.

A reduction in mycotoxins can be achieved at different levels, depending on which production phase is involved. The presence can be controlled by:

1. Prevention
2. Decontamination of the mycotoxin-content.

Without any doubt, prevention is the first step that should be undertaken to control the presence of mycotoxins. It has been demonstrated that Hazard Analysis Critical Control Points plans (HACCP) can be reliable, but they do not guarantee the elimination of mycotoxin-producing organisms or of the toxin itself (Gil et al., 2016; Coton et al., 2019; Finne Kure and Skaar, 2019; Sánchez-Montero et al., 2019a).

Generally speaking, decontamination can be achieved through different physical treatments, such as: heating, oxidizing agents, microwaves or radiation (Massoud et al., 2018).

Heat has been proved effective when the temperature is above 100°C and, in addition, the presence of NaOH and the moisture content seem to enhance the effect, even if not completely. This treatment can be useful for food of vegetal origin (bakery products, coffee beans and cocoa) and of animal origin (pig products) but with a lower effect. Different authors reported a reduction of as much as 93.6% for a treatment of cocoa beans at 110°-140°C for 30 minutes and up to 97% for coffee beans, but a reduction of only 20% for pig products after a frying process (Santini et al., 2015; Karlowsky et al., 2016). An ozone treatment at 10% for 15 s was able to degrade OTA in an *in vitro* assay. Microwaves can be more efficient. Radiation, even though more effective, poses other problems in terms of resistant microorganisms, the nutritional values of radiated food and consumers' acceptability (Massoud et al., 2018).

Chemical preservatives or, in a modified atmosphere, packaging could be applied to control the incidence of mycotoxins in meat products. However, these treatments are not appropriate for dry-cured meat products, since fungus activity is essential for their sensory characteristics (Acosta et al., 2009; Schmidt-Heydt et al., 2013; Bernáldez et al., 2014; Núñez et al., 2015; Rodríguez et al., 2015a, 2015b). Consumers are currently demanding products that are free of chemical residues, and as a result the use of chemical fungicides that can leave residues in meat cannot be adopted (Núñez et al., 2015). The processing of industrial meat, involving heating, salting, drying and storage, is ineffective in reducing OTA concentrations in the final product, as pointed out by different authors (Perši et al., 2014; Domijan et al., 2015; Pleadin et al., 2015).

Moreover, considering that OTA represents a moderately stable molecule, which is resistant to high temperatures and capable of surviving in most food processing stages, (i.e., boiling, cooking, frying, roasting, fermentation), the risk of OTA contamination in the final meat products increases considerably (Domijan et al., 2015; Pleadin et al., 2015). A 20% reduction in mycotoxin content, obtained utilizing some of these meat cooking procedures, has been achieved (Perši et al., 2014).

Ozone, the triatomic form of oxygen (O_3), and one of the most powerful disinfectants to have been approved for direct application as an antimicrobial agent in the food industry (Udomkun et al., 2017), is another method that may be used to reduce mycotoxin-producing fungi in meat products.

Gamma radiation (γ) has also been considered as a tool for the preservation and maintenance of food quality (Udomkun et al., 2017). It has been found to be a simple and efficient decontamination technique, since it successfully destroys the microorganisms that cause deterioration and reduces some toxins, such as mycotoxins, that compromise the quality of food in terms of its nutritional and sensory properties. The efficiency of γ radiation depends on many factors, such as the number and type of fungal lineages, the dose, food composition and air humidity (Aquino, 2012; Jalili et al., 2012).

Only limited studies on the effects of γ-radiation on OTA levels are available, and the results of these studies are contradictory because some studies showed that γ radiation can reduce OTA levels, even if applied at low levels (doses), while others have stated that a reduction in OTA can only be achieved for higher irradiation doses (Domijan et al., 2015). A reduction in OTA-producing fungi through chemical methods has already been described by some authors. Both a partial and a total inhibition of growth of these microorganisms, through the use of NaCl, were observed for dry-cured ham (Schmidt-Heydt et al., 2011).

Biopreservation or biocontrol is intended as "the use of natural or added microorganisms, fermentates or their metabolites to increase food safety". This option has been attempted for Lactic Acid Bacteria (LAB), but more microbes, yeast and their mechanisms have been studied over the last few decades (Camardo Leggieri et al., 2017; Leyva Salas et al., 2017; Sheikh-Zeinoddin and Khalesi, 2019). This option has two main advantages: it removes the need to use chemical or physical treatments that can be detrimental for the beneficial molds and, being a mild process, it enhances the aroma compounds and the retention of nutrients. For all these reasons, biopreservation can be considered a very valuable strategy for food of animal origin (Finne Kure and Skaar, 2019; Sheikh-Zeinoddin and Khalesi, 2019).

Unfortunately, it is not such an immediate choice because some basic criteria must be met, as stated in the past (Karlowsky et al., 2016; Darwish, 2019; Sheikh-Zeinoddin and Khalesi, 2019):

1. Mycotoxins must be inactivated or destroyed as non toxic compounds in an irreversible way;
2. Fungal spores and mycelia should be destroyed (no new toxin production);
3. The food should still be palatable and retain its nutritive value;
4. The use and the treatment should be economically feasible (final value of the commodity).

Biological decontamination can be achieved in different ways: by competition with toxigenic mold for different substrates, by modulation of the OTA production, by degradation or by binding of the OTA (Siedler et al., 2019).

The competition mechanism has recently been studied with Proteomic techniques. Several proteins, for example, the Grg-1 protein (a glucose-repressible protein) or the catabolite repression protein CreC, of *P. chrysogenum*, or of *D. hansenii*, and enzymes have demonstrated competition in the glucose pathway. The same study also suggested that these microorganisms can also inhibit the growth of OTA producing strains by impairing the wall integrity (Delgado et al., 2019).

Other microorganisms and fungi that are able to compete with OTA producing strains are: *L. plantarum* CRL 778 (Dallagnol et al., 2019), *L. brevis* KR3 and KR51 in fermented milk products (Tropcheva et al., 2014), *P. nalgiovense* in dairy products (Camardo Leggieri et al., 2017), and *S. fibuligera*, *Candida zeylanoides* and *Hyphopichia burtoni* in meat products (Iacumin et al., 2017; Meftah et al., 2020).

Counteracting the production of OTA – a modulation mechanism - is another possibility. Several authors have highlighted the capacity of inhibition of the expression of the *otanpsPN* gene, a specific gene associated with OTA production, by specific bacterial strains of *D. hansenii* (Merla et al., 2018; Perrone et al., 2019). It should be noted that this strategy can also be applied for the control of other mycotoxins, such as aflatoxins.

Degradation can be obtained via several fermentation processes through the production of acid or alcoholic compounds. Lactic or alcoholic fermentation is involved and lactic fermentation can easily be achieved in dairy products (Perrone et al., 2019). Lactic Acid Bacteria (LAB) produce hydrogen peroxide (H_2O_2), which is then converted into hypothiocyanate via the reaction of hydrogen peroxide and thyiocyanate, a reactive oxygen species, with the aid of lactoperoxidase. These oxygen reaction species may affect mycotoxin production. LAB, depending on the species and strain, have shown the ability to control OTA production (Bagher Hashemi and Gholamhosseinpour, 2019; Chiocchetti et al., 2019).

Enzymes can also degrade OTA. Peptidases in particular have been found to be reliable in hydrolyzing OTA, because it is an amide (Karlowsky et al., 2016). The first enzymes discovered were Carboxypeptidase A (CPA), which derived from the pancreas and the α-chymotripsin, which belongs to the carbossipetidase family. Other enzymes, such as lipases, and esterases have also been found to be involved in OTA degradation (Karlowsky et al., 2016; Sheikh-Zeinoddin and Khalesi, 2019). Enzymes can be used as bacterial cultures, but also as cell-free supernatant cultures or centrifugated preparations where enzymes are collected. However, cell-free and centrifugated preparations have less activity. Bagher Hashemi and Gholamhosseinpour (2019) concluded that this behavior is probably due to the aggregation of high molecular active compounds during the centrifugation and filtration preparation steps. A recent study has on the contrary highlighted that this behavior is not only a question of preparation, but that it also depends on which OTA producing mold species are involved. Different forms of *Candida zeylanoides* (live cells, cell broth or diffused compounds) can act in opposite ways, depending on the different producing molds. Live cells and cell broth are known to have inhibited toxin production in the presence of *P. nordicum* but also to have stimulated toxin production in the presence of *A. westerdijakie* (Meftah et al., 2020).

Several enzyme products are already available on the market (Karlowsky et al., 2016; Sheikh-Zeinoddin and Khalesi, 2019). Genetic engineering and the biotechnology of OGM microorganisms in the past and more recently have offered new possibilities for commercial preparations (Chang et al., 2015; Massoud et al., 2018). For example, Chang et al. (2015) studied the use of in vitro engineered microorganisms using a carbopetidase derived from *Bacillus amyloliquefaciens* ASAG1, isolated from maize, on recombinant *E. coli* cells.

Competition, toxin production control and degradation can interact and have different weights in a decontamination process, as pointed out for the case of different *L. plantarum* strains (PTCC1058, LP3, AF1 and LU5), where competition for substrates, the production of acids and inhibition by antagonistics compounds have been confirmed (Bagher Hashemi and Gholamhosseinpour, 2019). These phenomena has also been observed for other spoilage mycetes (Siedler et al., 2019).

In addition to their antifugal activity, bacterial strains can be used because they have other positive effects on food. This is the case of two strains of *L. plantarum* (RG7B and C11C) and one strain of *Pediococcus pentosaceus* (RG8A), isolated from matrices, other than food, of animal origin. The ability of these strains to resist low pH environments and to tolerate 0.3% bile salts also suggests their use as probiotics (Taroub et al.,

2019). Other species suggested as biocontrol agents alone or in association have been studied, and the authors have also suggested their use as probiotics: *Bifidobacterium animalis* sub. *lactis* (strain A026), *L. acidophilus, L. rhamnosus* (strains A238, A119 (Leyva Salas et al., 2017; Sheikh-Zeinoddin and Khalesi, 2019).

Another mechanism that is used for the biocontrol of food is the binding of OTA by different yeast and LAB components. The glucans of the cell wall are involved in this process (Chiocchetti et al., 2019). *L. acidophilus* and *L. plantarum,* used together with prebiotics, have shown a removal efficiency of over 85%. The same mechanism seems to be involved when *S. cerevisiae* is used. In this case, a heat treatment seems to enhance the binding capacity, probably as a result of the formation of Maillard reaction products, protein denaturation, thinning of the cell wall and enlargement of the wall pores. All these effects lead to a greater exposition to binging sites (Chen et al., 2018). Unfortunately, the binding option is only useful for certain liquid foods, such as wine and beer, but not for meat or dairy products, where bacterial cells, bearing the bind OTA, cannot be separated after treatment (Massoud et al., 2018).

Whatever the mechanism of action is, the type of product and its production process should be taken into account to obtain the maximum result and the concentration of controlling starters. Iacumin et al. (2017), for example, when dealing with speck, an Italian cured meat product, suggested inoculating the yeast culture after the smoking phase and before the drying and seasoning steps at a concentration > 10^6 UFC/cm^2 in order to ensure the maximum growth and the dominating effect over OTA producing molds. Another important factor is acidity. Neutral pH is the optimum level, even though the authors suggested using a more acid environment due to the fact that many foods exposed to OTA contamination have a low pH. Other factors that should be considered are: the incubation period and the temperature, which not always coincide with the optimal development temperature of the biocontrolling bacterial or yeast strain involved (Sheikh-Zeinoddin and Khalesi, 2019).

Not only can microorganism be involved in biocontrol, and several studies have in fact proposed the use of specific spices in food preparations as essential oils or directly as ingredients. However, their mechanisms are still controversial. Several hypotheses have been put forward (Ozcakmak et al., 2017; Delgado et al., 2019; Sánchez-Montero et al., 2019b):

- interaction with the biomembrane through the hydrophobic benxene ring with aliphatic side chains;
- formation of hydrogen bonds by hydroxyl groups;
- acidification by phenolic compounds;
- inhibition of cellular respiration and changes in the membrane that can lead to a loss of homeostasis;
- block of the enzymatic energy and mycotoxin production systems;
- interaction with membrane proteins which can result in a loss of protons throughout the membrane itself.

Among the substances tested so far, the most reliable against the growth and production of toxins are wild oregano, due to its carvacol and tymol compounds, and garlic, due to its allyl sulfide and allyl disulfide compounds, which should be used at concentrations of 250 and 500 μL/mL till the eighth day. Sage, due to its 1,8-cinole, borenol, tyranton and camphor contents, and mint, due to its menthone, menthol and neomenthole contents, are effective, albeit to a lesser extent, and should be used for fewer days (4 days) (Sánchez-Montero et al., 2019b). Smoked paprika, which is used in the Spanish "*Pimenton de La Vera*" spice, was found to be able to inhibit toxin production and increase the lag phase of the mold growth curve at a concentration of 2% and a_w 0,98- 0.94. These effects have been attributed to: capsaicin, capsanthin, capsorubin, saponins and the aforementioned fenolic compounds (Kollia et al., 2019; Sánchez-Montero et al., 2019b). Flavonoids also seem to be effective in controlling OTA production (Ricelli et al., 2019).

The last approach mentioned here for the control of OTA production is the use of substances that can prevent mold growth on the surface of dairy and meat products. Several studies have been conducted with chitosan coating systems. Chitosan is the deacetylated form of chitin, a natular component of crustacean shells and fungal wall cells. Its properties (biodegradability, biocompatibility, non-toxicity) have made this substance a natural and Generally Recognized As Safe (GRAS) food additive. The fungistatic effect is linked to the polisaccaride can interfer with minerals uptake (Ca^{2+}) and nutrients delaying the germination (De Elguea-Culebras et al., 2019). Chitosan, alone or used as a potential carrier of such by-products as those seen in the previous part (can be used for coating purposes and have been proved to be reliable in cheese and meat fermented products (Arslan and Soyer, 2018; De Elguea-Culebras et al., 2019).

The action of all these substances is limited in time and is dose dependent, but they can still be useful. They can be used together with other parameters or under other conditions while awaiting the development of protective cultures in accordance with the "hurdle theory" (Sánchez-Montero et al., 2019b).

By controlling the a_w level and adjusting the drying and ripening temperatures, it is possible to reduce the production and accumulation of toxic secondary toxic metabolites since the a_w of the substrate affects the production capacity of fungal mycotoxins. A significantly higher verrucosidine production by *Penicillium polonicum* was reported for an a_w of 0.99, compared to a_w of 0.97 and 0.95 (Asefa et al., 2011). Biological methods, based on the competitive exclusion of non-toxigenic fungal strains, have been reported as a promising approach to mitigate the formation of mycotoxins and prevent their absorption in the human body (Udomkun et al., 2017). Among these biological methods, the biocontrol of toxigenic fungi, by means of antagonistic microorganisms, is an alternative to chemical and physical methods (Núñez et al., 2015). A valuable strategy to prevent the growth of ochratoxigenic molds in dry-cured meat products is the use of native yeasts and non-toxigenic fungi as biopreservative agents (Virgili et al., 2012; Bernáldez et al., 2013; Andrade et al., 2014; Simoncini et al., 2014; Rodríguez et al., 2015b).

Microorganisms that are able to produce active compounds against unwanted molds, such as antifungal proteins, with potent activity against toxigenic species, are commonly found in dry-cured meats (Rodríguez et al., 2015a). Thus, the biopreservation of food could be achieved through the use of antimicrobial proteins (small basic proteins) secreted from filamentous fungi. Antifungal proteins produced by several strains of the *Aspergillus* and *Penicillium* genera have been described (Bernáldez et al., 2014) and many proteins and peptides with antifungal activity from plants, bacteria, arthropods, amphibians or reptiles have been purified and characterized. Delgado et al., (2015) investigated antifungal proteins (AFP) from *Aspergillus giganteus*, Anafp from *Aspergillus niger*, AcAFP from *Aspergillus clavatus*, NFAP from *Neosartorya fischeri*, and PAF, PgAFP and Pc-Arctin from *Penicillium chrysogenum*. The antimicrobial activity of PgAFP against common microorganisms in dry foods and their sensitivity to proteolytic enzymes and heat treatments were evaluated by Ferrara et al. (2015), who concluded that this protein shows a potent inhibitory activity against such undesirable fungi as *Aspergillus* and *Penicillium*, the main mycotoxin producing species of concern for dry foods.

In a study carried out on inoculated dry-fermented salami, the PgAFP efficiently reduced the *Aspergillus flavus* and *Penicillium restrictum* counts. A *Penicillium chrysogenum* strain producing PgAFP was isolated from dried ham, and this protein is important for the control of dangerous fungi in dry-matured foods. Since PgAFP is known to be active against some toxigenic species of *Penicillium* and *Aspergillus*, the potential inhibition of other undesirable organisms of importance for dry food deserves further investigation (Delgado et al., 2015).

The *Penicillium chrysogenum* RP42C strain, in the presence of toxigenic fungal strains, demonstrated a high growth capacity under ripening conditions, and it could therefore be used as a protective crop in the ripening process of dried ham (Bernáldez et al., 2014).

In light of these considerations, OTA-producing molds could be controlled in the meat industry during the maturation of dry-cured Iberian ham through the use of protective non-toxigenic mold cultures (Rodríguez et al., 2015a). Other food characteristics, such as a_w, temperature, chemical composition and microbial population, may affect the efficacy of PgAFP (Delgado et al., 2015), thus, an adequate knowledge of the mode of action of antagonistic yeasts is useful to both improve its performance against toxigenic fungi and to establish screening criteria in order to obtain more effective strains. Another possible decontamination strategy is related to the use of yeasts, which may decrease the mycotoxin content through adsorption to such cell wall molecules as glycoproteins, or by blocking the biosynthetic pathway of mycotoxins (Núñez et al., 2015).

The yeast species, the level of fungus contamination and the presence of NaCl in the culture medium all influence the activity of PgAFP, and the efficacy of yeasts in the inhibition of *in vitro* fungal growth and OTA production. In fact, the growth of halophilic strains, such as *Debaryomyces hansenii,* is promoted by the NaCl content, with a consequent significant increase in biocontrol activity. *Hyphopichia burtonii* has demonstrated an *in vitro* biocontrol activity against *Penicillium nordicum* (Simoncini et al., 2014).

Debaryomyces hansenii, the predominant yeast species during the processing of dry-cured meat products, may be useful for the development of starter cultures to improve the safety and sensory quality of dry-cured meat products, because it is effective in reducing pathogenic fungi and also contributes to the development of the known flavor of these products (Andrade et al., 2014; Núñez et al., 2015). Similarly, *Debaryomyces hansenii* 253H and 226G isolates, which are able to colonize and prevail in dried fermented salami, have shown an antifungal effect. These two isolates may be considered for use as protective cultures against toxigenic molds in dried fermented meat products (Núñez et al., 2015).

References

Acosta, R., Rodríguez-Martín, A., Martín, A., Núñez, F., Asensio, M.A., 2009. Selection of antifungal protein-producing molds from dry-cured meat products. *International Journal of Food Microbiology* 135, 39–46. https://doi.org/10.1016/j.ijfoodmicro.2009.07.020.

Alapont, C., López-Mendoza, MC., Gil, JV., Martínez-Culebras, PV. 2014. Mycobiota and toxigenic *Penicillium* species on two Spanish drycured ham manufacturing plants. *Food Additives and Contaminants Part A* 31, 93–104. https://doi.org/10.1080/19440049.2013.849007.

Andrade, M.J., Thorsen, L., Rodríguez, A., Córdoba, J.J., Jespersen, L., 2014. Inhibition of ochratoxigenic moulds by *Debaryomyces hansenii* strains for biopreservation of drycured meat products. *International Journal of Food Microbiology* 170, 70–77. https://doi.org/10.1016/j.ijfoodmicro.2013.11.004.

Anelli, P., Haidukowski, M., Efifani, F., Cimmarusti, M.T., Moretti, A., Logrieco, A., Susca, A. 2019. Fungal mycobiota and mycotoxin risk for traditional rtisanal italian Cave Cheese. *Food Microbiology* 78:62-72 https://doi.org/10.1016/j.fm.2018.09.014.

Aquino, K.A.S., 2012. Sterilization by gamma irradiation. In: Adrovic, F. (Ed.), Gamma Radiation. *InTech,* Vienna, Austria, pp. 171–206.

Arslan, B. and Soyer, A. 2018. Effects of chitosan as urface fungus onhibitor on microbiological, physicochemical, oxidative and sensory characteristics of dry fermented sausages. *Meat Science* 145:107:113 https://doi.org/10.1016/j.meatsci.2018.06.012.

Asefa, D.T., Kure, C.F., Gjerde, R.O., Langsrud, S., Omer, M.K., Nesbakken, T., Skaar, I., 2011. A HACCP plan for mycotoxigenic hazards associated with dry-cured meat production processes. *Food Control* 22: 831–837. https://doi.org/10.1016/j.foodcont.2010.09.014.

Bagher Hashemi, S.M. and Gholamhosseinpour, A. 2019. Fermentation of Table Cream by *Lactobacillus plantarum* strains: effect on fungal growth, Aflatoxin M_1 and Ochratoxin A. *International Journal of Food Science and Technology* 54:347-353 doi:10.1111/ijfs.13943.

Becker-Algeri, T.A., Castagnaro, D., De Bortoli, K., De Souza, C., Drunkler, D.A., Badiale-Furlong, E. 2016. Mycotoxins in bovine milk and dairy products: a Review. *Journal of Food Science*, 81:R544-R552 doi: 10.1111/1750-3841.13204.

Bernáldez, V., Córdoba, J.J., Rodríguez, M., Cordero, M., Polo, L., Rodríguez, A., 2013. Effect of Penicillium nalgiovense as protective culture in processing of dry-fermented sausage "salchichón". *Food Control* 32: 69–76.

Bernáldez, V., Rodríguez, A., Martín, A., Lozano, D., Córdoba, J.J., 2014. Development of a multiplex qPCR method for simultaneous quantification in dry-cured ham of an antifungal-peptide Penicillium chrysogenum strain used as protective culture and aflatoxin-producing moulds. *Food Control* 36: 257–265. https://doi.org/10.1016/j.foodcont.2013.08.020.

Camardo Leggieri, M., Decontardi, S., Bertuzzi, T., Pietri, A., Battilani, P. 2017. Modeling growth and toxin production of toxigenic fungi signaled in cheese under different temperature and water activity regimes. *Toxins*, 9,4 doi:103390/toxins9010004.

Capozzo, J., Jackson, L., Lee, H.J., Zhou, W., Al-Taher, F., Zweigenbaum, J., Ryu, D. 2017. Occurrence of Ochratoxin A in infant foods in United States. *Journal of Food Protection* 80: 251-256 doi:10.4315/0362-028X.JFD-16-339.

Carballo, D., Tolosa, J., Ferrer, E., Berrada, H. 2019. Dietary exposure assessment to mycotoxins trought Total Diet Studies. a review. *Food and Chemical Toxicology* 128:8-20 https://doi.org/101016/j.fct.2019.03.033.

Chang, X., Wu, Z., Wu, S., Dai, Y., Sun, C 2015. Degradation of Ochratoxin A by *Bacillus amyloliquefaciens* ASAG1. *Food Additives and Contaminants,* Part A 32:564-571 http://dx.doi.irg/10.1080/19440049.2014.991948.

Chen, W., Li, C., Zhang, B., Zhou, Z., Shen, Y., Liao, X., Yang, J., Wang, Y., Li, X., Li, Y., Shen, X.L. 2018. Advances in biotoxification of Ochratoxin A - A review of the past five decades. *Frontiers in Microbiology*, 9:1386 doi:10.3389/fmic.2018.01386.

Chiocchetti, G.M., Jadán-Piedra, C., Monedero, V., Zúñiga, M., Vélez, d., devesa, V. 2019. Use of Lactic Acid bacteria and yeasts to reduce exposure to chemical food contaminants and toxicity. *Critical Reviews in Food Science and Nutrition* 59:1534-1545 https://doi.org/10.1080/10408398.2017.1421521.

Conte, F., Panebianco, A. 2019. Potential hazards associated with raw donkey milk consumption: a review. *International Journal of Food Science* Article ID 5782974, https://doi.org/10.1155/2019/5782974.

Coton, M., Auffret, A., Poirier, E., Debaets, S., Coton, E. 2019. Production and migration of Ochratoxin A and Citrin in Comté Cheese by an osolate of *Penicillium verrucosum* selected among *Penicillium* spp mycotoxin producers in YES Medium. *Food Microbiology*, 82:551-559 https://doi.org/10.1016/j.fm.2019.03.026.

Dallagnol, A.M., Bustos, A.Y., Martos, G.I., Font de Valdez, G.; Gerez, C.L. 2019. Antifungal and antimycotoxigenic effect on *Lactobacillus plantarum* CRL 778 at different Water Acivities values. *Revista Argentina de Microbiología* 51:164-169 https://doi.org//10.1016/j.ram.2018.04.004.

Dall'Asta, C., De Dea Linder, J., Galaverna, G., Dossena, A., Neviani, E., Marchelli, R. 2008. The Occurrence of Ochratoxin A in Blue Cheese. *Food Chemistry* 106:729-734 doi:10.1016/j.foodchem.2007.06.049.

Dall'Asta, C., Galaverna, G., Bertuzzi, T., Moseriti, A., Pietri, A., Dossena, A., Marchelli, R., 2010. Occurrence of ochratoxin A in raw ham muscle, salami and dry-cured ham from pigs fed with contaminated diet. *Food Chemistry* 120: 978–983. https://doi.org/10.1016/j.foodchem.2009.11.036.

Darwish, A.M.G. 2019. Fungal mycotoxins and natural antioxidants: two sides of the same coin and significance in food safety. *Microbial Biosystems* 4:1-16.

Decontardi, S., Mauro, A., Lima, N., Battilani, P. 2017. Survey of penicilla Associated with Italian Grana Cheese. *International Journal of Food Microbiology* 246:25-31 http://dx.doi.org/10.1016/j.ilfoodmicro.2017.01.019.

De Elguea-Culebras, G.O., Bourbon, A.I., Costa, M.J., Muñoz-Tebar, N., Carmona, M., Molina, A., Sánchez-Vioque, R., Beruga, M.I., Vicente, A.A. 2019. Optimization of a chitosan solution as potential carrier for the incorporation of *Santolina chamaecyparissus* L. solid by-product in a edible vegetal coating on '*Manchego*' cheese. *Food Hydrocolloids* 89:272-282 https://doi.org/10.1016/l.foodhyd.2018.10.054.

Delgado, J., Acosta, R., Rodríguez-Martín, A., Bernúdez, E., Núñez, F., Asensio, M.A., 2015. Growth inhibition and stability of PgAFP from *Penicillium chrysogenum* against fungi common on dry-ripened meat products. *International Journal of Food Microbiology* 205: 23–29. https://doi.org/10.1016/j.ijfoodmicro.2015.03.029.

Delgado, J.; Núñez, F., Asencio, M.A.; Owens, R.A. 2019. Quantitative Proteomic Profiling of Ochratoxin A repression in *Penicillium nordicum* by protective cultures. *International Journal of Food Microbiology* 305:108243 https://doi.org/10.1016/j.ijfoodmicro.2019.108243.

Domijan, A.M., Pleadin, J., Mihaljevié, B., Vahčić, N., Frece, J., Markov, K., 2015. Reduction of ochratoxin A in dry-cured meat products using gamma-irradiation. *Food Additive Contaminants Part. A.* 32: 1185–1191. https://doi.org/10.1080/19440049.2015.1049219.

EFSA (2006) European food safety authority. Opinion of the scientific panel on contaminants in the food chain on a request from the Commission related to ochratoxin A in food. *EFSA J* 365:1–56.

Elaridi, J., Dimassi, H., Fasan, H. 2019. Aflatoxin M1 and Ochratoxin A in baby formulae marketed in Lebanon: occurrence and safety evaluation. *Food Control* 106:106680 https://doi.org/10.1016/j.foodcont.2019.06.006.

Ferrara, M., Perrone, G., Gallo, A., Epifani, F., Visconti, A., Susca, A., 2015. Development of loop-mediated isothermal amplification (LAMP) assay for the rapid detection of *Penicillium nordicum* in dry-cured meat products. *International Journal of Food Microbiology* 202, 42–47. https://doi.org/10.1016/j.ijfoodmicro.2015.02.021.

Fink-Gremmels J. and Van Der Merwe D. 2019. Mycotoxins in the food chain: contamination of foods of animal origin. IN "Chemical Hazards in Food of Animal Origin", *Food safety Assurance and Veterinary Public Health no. 7*, edited by Smulders, J.M., Rietjens, I.M.C.M., Rose, M.D. Waeningen Academic Publisher DOI 10.3920/978-90-8686-877-3_10.

Finne Kure, C., and Skaar, I. 2019. The fungal problem in cheese industry. *Current Opinion in Food Science* 29:14-19 https://doi.org/10.1016/j.cofs.2019.07.003.

Flores-Flores, M.E., Lizarraga, E., López de Cerain, A., Gonzáles-Peñas, E. 2015 Presence of mycotoxins in animal milk: a review. *Food Control* 53:163-176 https://dx.doi.org/10.1016/j.foodcont.2015.01.020.

Flores-Flores, M.E. and Gonzáles-Peñas, E. 2018. Short Communication: Analysis of mycotoxins in spanish milk. *Journal of Dairy Science* 101:113-117 https://doi.org/10.3168//jds.2017-13290.

Gareis, M., Scheuer, R., 2000. Ochratoxin A in meat and meat products. *Archiv für Lebensmittelhygiene* 51:102–104.

Gil, L., Ruiz, P., Font, G., Manyes, L. 2016. An Overview of the application of Hazard Analysis and Critical Control Point (HACCP) system to Mycotoxins. *Reviews in Toxicology* 33:50-55.

Gross, M., Puck Ploetz, C., Gottschalk, C. 2019. Immunochemical detection of mycotoxins in donkey milk. *Mycotoxin Research* 35:83-87 https://doi.org/10.1007/s12550-018-0333-2.

Hof, H. 2016. Mycotoxins in milk for human nutrition: cow, sheep and human breast milk. *GMS Infectious Diseases* 4:Doc03 DOI: 10.3205/id000021.

Hymery, N., Vasseur, V., Coton, M., Mounier, J., Jany, J-L., Barbier, G., Coton, E. 2014. Filamentous fungi and mycotoxins in cheese: a review. *Comprehensive Reviews in Food Science and Food Safety* 13:437-456 doi: 10.1111/1541-4337.12069.

Iacumin, L., Manzano, M., Andyanto, D., Comi, G. 2017. Biocontrol of ochratoxigenic moulds (*Aspergillus ochraceus* and *Penicillium nordicum*) by *Debaryomices hansenii* and *Saccharomycopsis fibuligera* during speck production. *Food Microbiology* 62:188-195. http://dx.doi.org/10.1016/j.fm.2016.10.017.

Iacumin L, Chiesa L, Boscolo D, Manzano M, Cantoni C, Orlic S, Comi G. 2009. Moulds and Ochratoxin A on surfaces of artisanal and industrial dry sausages. *Food Microbiology* 26:65–70 http://dx.doi:10.1016/j.fm.2008.07.006.

Iqbal, S.Z., Nisar, S., Asi, M.R., Jinap, S., 2014. Natural incidence of Aflatoxins, Ochratoxin A and Zearalenone in chicken meat and eggs. *Food Control* 43: 98–103. https://doi.org/10.1016/j.foodcont.2014.02.046

Jalili, M., Jinap, S., Noranizan, A., 2012. Aflatoxins and Ochratoxin A reduction in black and white pepper by gamma radiation. *Radiation*

Physic Chemistry 81: 1786–1788. https://doi.org/10.1016/j.radphysch em.2012.06.001.

Kamal, R.M., Mansour, M.A., Elalfy, M.M., Abdelfatah, E.N., Galala, W.R. 2019. Quantitative detection of Aflatoxin M1, Ochratoxin and Zearalenone in fresh raw milk of cow, bufalo, sheep and goat by HPLC XEVO-TQ in Dakahlia Governatorate, Egypt. *Journal of Veterinary Medicine and Health* 3:114.

Karlowsky, P., Suman, M., Berthiller, F., De Meester, J., Eisenbrand, G., Perrin, I., Oswald, I.P., Speijers, G., Chiodini, A., Recker, T., Dussort, P. 2016. Impact of food processing and detoxification treatments on mycotoxin Contamination. *Mycotoxin Research* 32:179-205 DOI 10.1007/s12550-016-0257-7.

Khalesi M, Khatib N. 2011. The effects of different ecophysiological factors on Ochratoxin A production. *Environmental Toxicology and Pharmacology* 32:113–121. https://doi.org/10.1016/j.etap.2011.05.013.

Kollia, E., Chatìralampos, P., Zoumpoulakis, P., Markaki, P. 2019. Capsaicin, an inihbitor of Ochratoxin A production by *Aspergillus* section *nigri* strains in grapes (*Vitis vinifera* L.). *Food Additives and Contaminants*, Part A 36:1709-1721 https://doi.org/10.1080/194400 49.2019.1652771.

Leyva Salas, M., Mounier, J., Valence-Bertel, F., Coton, M., Thierry, A., Coton, E. 2017. Antifungal microbial agents for food biopreservation – a review. *Microrganisms*, MDPI, 5 Open Access 10.3390/microrg anisms5030037. Hal-01568139.

Lippolis, V., Ferrara, M., Cervellieri, S., Damascelli, A., Epifani, F., Pascale, M., Perrone, G., 2016. Rapid prediction of Ochratoxin A - producing strains of Penicillium on drycured meat by MOS-based electronic nose. *International Journal of Food Microbiology* 218: 71–77. https://doi.org/10.1016/j.ijfoodmicro.2015.11.011.

Luo, X., Liu, X., Li, J. 2018. Updating techniques on controlling mycotoxins – A Review. *Food Control* 89:123-132. https://doi.org/10.1016/j.fo odcont.2018.01.016.

Mao, J., Zheng, N., Wen, F., Guo, L., Fu, C., Ouyang, H., Zhong. L., wang, J., Lei, S. 2018. Multi-mycotoxin analysis in raw milk by Ultra High

Performance Liquid Chromatogralhy coupled to Quadrupole Orbitrap Mass Spectrometry. *Food Control* 84:305-311 https://dx.doi.org/ 10.1016/j.foodcont.2017.08.009.

Markov, K., Pleadin, J., Bevardi, M., Vahčić, N., Sokolić-Mihalak, D., Frece, J., 2013. Natural occurrence of Aflatoxin B1, Ochratoxin A and Citrinin in croatian fermented meat products. *Food Control* 34, 312–317. http://dx.doi.org/10.1016/j.foodcont. 2013.05.002.

Massoud, R., Cruz, A., Darani, K.K. 2018. Ochratoxin A: from Safety Aspects to prevention and remediation strategies. *Current Nutrition & Food Science*, 14:11-16; https://doi.org/10.2174/157340131366617 0517165500.

Meftah, S., Abid, S., Dias, T., Rodrigues, P. 2020. Mechanism underlying the effect of commercial starter cultures and a native yeast on Ochratoxin A production in meat products. *Food Science and Technology* 117:108611 https://doi.org/10.1016/j.lwt.2019.108611.

Merla, C., Andreoli, G., Garino, C., Vicari, N., Tosi, G., Guglielminetti, M.L., Moretti, A., Biancardi, A., Arlorio, M., Fabbi, M. 2018. Monitoring of Ochratoxin A and ochratoxin producing fungi in traditional salami manufactured in northern Italy. *Mycotoxin Research*, 34:107-116 https://doi.org/10.1007/s12550-017-0305-y.

Milicevic, D., Nesic, K., Jaksic, S. 2015. Mycotoxin contamination of the food supply chain – Implications for the One Health programme. *Procedia Food Science* 5: 187-190 doi: 10.1016/j.profoo.2015.09.053.

Montanha, F.P., Anater, A., Burchard, J.F., Luciano, F.B., Meca, G., Manyes, L., & Pimpao, C.T. (2018). Mycotoxins in dry-cured meats: a review. *Food and Chemical Toxicology* 111, 494-502. https://doi. org/10.1016/j.fct.2017.12.008.

Núñez, F., Lara, M.S., Peromingo, B., Delgado, J., Sanchez-Montero, L., Andrade, M.J., 2015. Selection and evaluation of *Debaryomyces hansenii* isolates as potential bioprotective agents against toxigenic penicillia in dry-fermented sausages. *Food Microbiology* 46: 114–120. https://doi.org/10.1016/j.fm.2014.07.019.

Olsen, M., Lindqvist, R., Bakeeva, A., Leong, S.L., Sulyok, M. 2019. Distribution of mycotoxins produced by *Penicillium* spp. onoculated in apple jam and Crème Fraiche during chilled storage. *International Journal of Food Microbiology* 292:13-20 https://doi.org/10.1016/j.ijfoodmicro.2018.12.003.

Ostry, V., Malir, F., Toman, J., Grosse, Y., 2016. Mycotoxins as human carcinogens-the IARC monographs classification. *Mycotoxin Research* 1–9. http://dx.doi.org/10.1007/s12550-016-0265-7.

Ozcakmak, S., Gul, O., Dervisoglu, M., Yilmaz, A., Sagdic, O., Arici, M. 2017. Comparison on the effect of some essential oils on the growth of *Penicillium verrucosum* and its ochratoxin Production. *Journal of Processing and Preservation* 41:e13006 doi: 10.1111/jfpp.134006.

Perrone, G., Rodriguez, A., Magistà, D., Magan, N. 2019. Insights into existing and future fungal and mycotoxin contamination of cured meats. *Current Opinion in Food Science*, 29: 20-27.

Perši, N., Pleadin, J., Kovačević, D., Scortichini, G., Milone, S., 2014. Ochratoxin A in raw materials and cooked meat products made from OTA-treated pigs. *Meat Science* 96: 203–210. https://doi.org/10.1016/j.meatsci.2013.07.005.

Pleadin, J., Staver, M.M., Vahčić, N., Kovačević, D., Milone, S., Saftić, L., Scortichini, G., 2015. Survey of Aflatoxin B1 and Ochratoxin A occurrence in traditional meat products coming from croatian households and markets. *Food Control* 52: 71–77. https://doi.org/10.1016/j.foodcont.2014.12.027.

Ramos-Pereira, J., Mareze, J., Patrinou, E., Santos, JA., López-Díaz, TM. 2019. Polyphasic identification of *Penicillium* spp isolated from spanish semi-hard ripened cheeses. *Food Microbiology* 84:103253 https://doi.org/10.1016/j.fm.2019.103253.

Ricelli, A., De Angelis. M., Primitivo, L., Righi, g., Sappino, G., C., Antonioletti, R. 2019. Role of some food-grade syntetized flavonoids on the control of Ochratoxin A in *Aspergillus carbonarius*. *Molecules* 24:2553 doi: 10.3390/molecules24142553.

Rodríguez, A., Medina, A., Córdoba, J.J., Magan, N., 2014. The influence of salt (NaCl) on ochratoxin A biosynthetic genes, growth and ochratoxin A production by three strains of *Penicillium nordicum* on a dry-cured ham-based medium. *International Journal of Food*

Microbiology 178: 113–119. https://doi.org/10.1016/j.ijfood micro.20b 14.03.007.

Rodríguez, A., Capela, D., Medina, Á., Córdoba, J.J., Magan, N., 2015a. Relationship between ecophysiological factors, growth and ochratoxin A contamination of dry cured sausage based matrices. *International Journal of Food Microbiology* 194: 71–77. https://doi.org/10.1016/j.ijfoodmicro.2014.11.014

Rodríguez, A., Bernáldez, V., Rodríguez, M., Andrade, M.J., Núñez, F., Córdoba, J.J., 2015b. Effect of selected protective cultures on ochratoxin A accumulation in drycured Iberian ham during its ripening process. *LWT-Food Science Technology* 60: 923–928. https://doi.org/10.1016/j.lwt.2014.09.059

Sánchez-Montero, L., Córdoba, J.J., Perodomingo, B., Alvarez, M., Núñez, F. 2019a. Effects on environmental conditions and substrate on growth and Ochratoxin A production by *Penicillium verrucosum* and *Penicillium nordicum*: relative risk assesment of OTA in dry-cured meat products. *Food Research International* 121:604-611 https:/doi.org/j.foodres.2018.12.025.

Sánchez-Montero, L., Córdoba, J.J., Perodomingo, B., Núñez, F. 2019b. Effect of spanish smoked paprika "Pimentón de La Vera" on control of Ochratoxin A and Aflatoxins production on a dry-cured meat model system. *International Journal of Food Microbiology* 308:108-303 https:/doi.org/j.ijfoodmicro.2019.108303.

Santini, A., Raiola, A., Meca, G., Ritieni, A. 2015. Aflatoxins, Ochratoxins, Trichothecenes, Patulin, Fumonisins and Beauvericin in finished products for human consumption. *Journal of Clinical Toxicology* 5:1000265 doi: 10.4172/2016-0495:1000265.

Schmidt-Heydt, M., Graf, E., Batzler, J., Geisen, R., 2011. The application of transcriptomics to understand the ecological reasons of Ochratoxin A biosynthesis by *Penicillium nordicum* on sodium chloride rich dry cured foods. *Trends Food Science Technology* 22, S39–S48. http://dx.doi.org/10.1016/j.tifs.2011.02.010.

Schmidt-Heydt, M., Stoll, D.A., Geisen, R., 2013. Fungicides effectively used for growth inhibition of several fungi could induce mycotoxin biosynthesis in toxigenic species. *International Journal of Food Microbiology* 166: 407–412. https://doi.org/10.1016/j.ijfoodmicro.2013.07.019.

Sengling Cebin Coppa, C.F., Khaneghah, A.M., Alvito, P., Assuncão, R., Martins, C., Eş, I., Gonçalves, B.L., Valganon de Neefi, D., Sant'Ana, A.S., Corassin, C.H., Fernandes Oliveira, C.A. 2019. The occurrence of mycotoxins in breast milk, fruit products and cereal-based infant formula: a review. *Trends in Food Science & Technology* 92:81-93 https://doi.org/10.1016/j.tifs.2019.08.014.

Sheikh-Zeinoddin, M. and Khalesi, M. 2019. Biological detoxification of Ochratoxin A on plants and plants products. *Toxin Reviews* 38:187-199 https://doi.org/10.1080/15569543.2018.1452264.

Siedler, S., Balti, R., Neves, A.R. 2019. Bioprotective mechanism of Lactic Acid Bacteria against fungal spoilage of food. *Current Opinion in Biotechnology* 56:138-146 https://doi.org/10.1016/j.copbio.2018.11.015.

Simoncini, N., Virgili, R., Spadola, G., Battilani, P., 2014. Autochthonous yeasts as potential biocontrol agents in dry-cured meat products. *Food Control* 46, 160–167. https://doi.org/10.1016/j.foodcont.2014.04.030.

Taheur, F.B., Kouidhi, B., Al Qurashi, Y.M.A., Salah-Abbès, J.B. 2019. Review: biotechnology of mycotoxin detoxification using microrganisms and enzymes. *Toxicon* 160:12-22 https://doi.org/10.1016/j.toxicon.201902.001.

Tao, Y., Xie, S., Xu, F., Liu, A., Wang, Y., Dongmei, C., Pan, Y., Huang, L., Peng, D., Wang, X., Yauan, Z. 2018. Ochratoxin A: toxicity, oxidative stress and metabolism. *Food and Chemical Toxicology* 112:320-331 https://doi.org/10.1016/j.fct.2018.01.002.

Taroub, B., Salma, L., Manel, Z., Ouzari, H-I., Hambi, Z., Moktar, H. 2019. Isolation of Lactic Acid Bacteria from grape fruit: antifungal activities, probiotic properties, and *in vitro* detoxification of Ochratoxin A. *Annals of Microbiology* 69:17-27 https://doi.org/10.1007/s132113-018-1359-6.

Tropcheva, R., Nikolova, D., Evstatieva, Y., Danova, S. 2014. Antifungal activity and identification of *Lactobacilli*, isolated from traditional dairy

Product "Katak". *Anaerobe* 28:78-84 http://dx.doi.org/10.1016/anaerobe2014.05.010.

Turkoglu, C. and Keyvan, E. 2019. Determination of Aflatoxin M_1 and Ochratoxin A in raw, pasteurized and UHT milk in Turkey. *Acta Scientiae Veterinariae* 47:1626 DOI: 10.22456/1679-9216.89667.

Udomkun, P., Wiredu, A.N., Nagle, M., Műller, J., Vanlauwe, B., Bandyopadhyay, R., 2017. Innovative technologies to manage aflatoxins in foods and feeds and the profitability of applications – a review. *Food Control* 76: 127–138. https://doi.org/10.1016/j.foodcont.2017.01.008.

Vipotnik, Z., Rodríguez, A., Rodrigues, P., 2017. *Aspergillus westerdijkiae* as a major Ochratoxin A risk in dry-cured ham based-media. *International Journal of Food Microbiology* 241, 244–251. http://dx.doi.org/10.1016/j.ijfoodmicro.2016.10.031.

Virgili, R., Simoncini, N., Toscani, T., Leggieri, M.C., Formenti, S., Battilani, P., 2012. Biocontrol of *Penicillium nordicum* growth and ochratoxin A production by native yeasts of dry cured ham. *Toxins* 4: 68–82. https://doi.org/10.3390/toxins4020068.

Volkel, I., Schroer-Merker, E., Czerny, C.P., 2011. The carry-over of mycotoxins in products of animal origin with special regard to its implications for the European Food Safety Legislation. *Food Nutrition Science* 2, 852–867. DOI: 10.4236/fns.2011.28117.

Younis, G., Ibrahim, D., Awad, A., El Bardisy, M.M. 2016. Determination of Aflatoxin M1 and Ochratoxin A in milk and dairy products in supermarkets located in Mansoura City, Egypt. *Advances in Animal and Veterinary Sciences* 4:114121 https://dx.doi.org/10.14737/journal.aavs/2016/4.2.114.121.

In: Ochratoxin A and Aflatoxin B1
Editor: Reuben Hess
ISBN: 978-1-53617-416-8

Chapter 2

OCHRATOXIN A CONTAMINATION OF TRADITIONAL DRY-CURED MEAT PRODUCTS

Jelka Pleadin[1,*], Manuela Zadravec[2], Tina Lešić[1], Ana Vulić[1], Nada Vahčić[3], Nina Kudumija[1], Jadranka Frece[3] and Ksenija Markov[3]

[1]Laboratory for Analytical Chemistry, Croatian Veterinary Institute, Zagreb, Croatia

[2]Laboratory for Feed Microbiology, Croatian Veterinary Institute, Zagreb, Croatia

[3]Faculty of Food Technology and Biotechnology, University of Zagreb, Zagreb, Croatia

* Corresponding Author's Email: pleadin@veinst.hr.

ABSTRACT

Dry-cured meat products produced using traditional technological processes represent food brands best recognised worldwide. During ripening, the surface of these products becomes overgrown with moulds whose spores mostly come from the environment in which ripening chambers are placed. The intensity of the overgrowth is enhanced by ripening longevity and traditional production environment in which, usually, no microbiological filters and no pneumatic barriers are used, so that the air temperature and relative humidity are virtually uncontrollable. Surface moulds are mainly of the *Penicillium* and *Aspergillus* genera, some of those having beneficial effects on product quality, and some causing mycotoxin contamination of the final products. The presence of the mycotoxin ochratoxin A (OTA) in dry-cured meat products can be mould-generated or come as a consequence of an indirect contamination (*carryover* effect) in case the meat-providing animals had been exposed to contaminated diet during farming. Another path of contamination are contaminated ingredients (e.g., spices) used in dry-cured meat products' production. This study investigated into the occurrence of OTA in different types of dry-fermented sausages and hams (n = 135) produced by a large number of households situated in different Croatian regions and sampled from the markets. Surface moulds were isolated and identified using traditional mycological and molecular (polymerase chain reaction, PCR) methods. The presence of OTA was first determined using a validated immunoassay method (ELISA); in samples with OTA concentrations higher than the limit of detection, the mycotoxin presence was confirmed using liquid chromatography tandem mass spectrometry (LC-MS/MS). OTA presence was determined in 11% of samples in the maximal concentration of 2.54 µg/kg in a ham sample. From the products' surfaces, a total of 404 *Penicillium*, 102 *Aspergillus* and 15 *Mucor* isolates were retrieved, out of which *Aspergillus niger, Aspergillus ochraceus* and *Penicillium verrucosum,* known as OTA- producers. The results showed random OTA contamination of traditional dry-cured meat products, indicating that, in order to avoid such a contamination, household-based meat & meat products' production should run under standardized and well-controlled conditions.

Keywords: traditional dry-cured meat products, moulds, mycotoxins, ochratoxin A

Introduction

Ochratoxin A (OTA) is considered to be one of the mycotoxins of the outermost importance from the public health standpoint. Coming as a result of contaminated food consumption, this mycotoxin can accumulate in the human body causing severe health impairments (Hussein and Brasel 2001; Richard 2007; Duarte et al. 2010; Marin et al. 2013). OTA has nephrotoxic, neurotoxic, mutagenic, carcinogenic, teratogenic and immunosuppressive effects in humans and animals (JECFA 2001); due to its evidenced toxic properties, the International Agency for Research on Cancer classified this mycotoxin into Group 2B of possible human carcinogens (IARC 1993). This mycotoxin can contaminate different types of food of animal origin, but its occurrence is rather characteristic for meat and meat products. Research on OTA in meat products is of particular importance, especially given that thermal processing, salting, drying and ripening, as well as different storage practices, have no significant impact on its level in final products (Bullerman and Bianchini 2007; Amézqueta et al. 2009; Pleadin et al. 2014).

OTA can be produced by various mould species of the *Aspergillus* and the *Penicillium* genus (Iacumin et al. 2009; Duarte et al. 2010), which overgrow the surface of dry-cured meat products. Numerous studies have demonstrated that, under certain circumstances such as favourable temperature, pH-value, water activity, presence or absence of crust or cracks, as well as with insufficient washing and brushing of dry-cured meats surface, i.e., in case of an uncontrolled mould growth, superficial moulds of the *Penicillium* and the *Aspergillus* genera produce OTA (Iacumin et al. 2009; Asefa et al. 2011; Rodríguez et al. 2012a; Rodríguez et al. 2015). It is important to mention that OTA present in meat products may also originate from farm animal feed, should the animals be fed with on feed contaminated with this mycotoxin (the *carry-over effect*), or can originate from spices used along the production line (Gareis and Scheuer 2000; Bertuzzi et al. 2013; Pleadin et al. 2013; Perši et al. 2014).

On top of the above, superficial moulds affect the quality of different dry-cured meat product and contribute to the development of product-specific flavour and taste (Asefa et al. 2009; Comi and Iacumin 2013;

Pleadin et al. 2017). That is the result of active participation of mould enzymes in fermentation and ripening, either by own virtue or in synergy with endogenous enzymes present in stuffing. A favourable impact on a meat product quality is also attributed to moulds' ability to retain moisture, which, in turn, prevents drying of the product's surface and incrustation that might arise due to protein coagulation (Toldrá 1998; Ockerman et al. 2000; Bruna et al. 2003). Due to their vast presence in production environments, the most commonly encountered moulds are those of the *Penicillium* genus (e.g., *P. chrysogenum*, *P. nalgiovense,* and *P. aurantiogriseum*) (Andersen 1995; Battilani et al. 2007; Sørensen et al. 2008; Sonjak et al. 2011). However, contrary to a favourable impact, certain moulds can have negative effects on the quality and safety of final meat products, since they produce mycotoxins as their secondary metabolites, including OTA as one of the most important mycotoxin contaminating this type of foodstuffs (Comi et al. 2004; Iacumin et al. 2009; Iacumin et al. 2011; Asefa et al. 2010).

It is known that spores of mycotoxin-producing moulds present on the surface of dry-cured meat products mostly come from the environment in which the ripening chambers are placed. The intensity of the overgrowth is enhanced by ripening longevity and traditional production environment in which usually no microbiological filters and no pneumatic barriers are used, and in which temperature and relative air humidity are virtually uncontrollable. It is not uncommon for OTA to be present in dry-cured meat products in substantial concentrations, especially in products with the longest ripening, such as hams. Investigations have also revealed that the outer casing cracking and skinlessness of certain hams facilitate OTA diffusion into the final product interior (Dall'Asta et al. 2010; Pleadin et al. 2015a; Pleadin et al. 2015b). In the colder geographical regions, OTA producers are mainly moulds of the *Penicillium* genus, most commonly *Penicillium verrucosum* and *Penicillium nordicum*, while in warmer regions the dominating moulds are those of the *Aspergillus* genus, most commonly *Aspergillus ochraceus* (Sonjak et al. 2011; Comi and Iacumin 2013).

As of now, no Maximum Levels (ML) for mycotoxins in meat and meat products have been stipulated by regulatory bodies (EC 2006a), making the data on OTA prevalence virtually unavailable. In order to prevent adverse consequences for human health, production of dry-cured meat products has to be based on the principles of good agricultural and manufacturing practices and Hazard Analysis Critical Control Point (HACCP) system, risk analysis, prevention and control of critical points, and obviation of possible contamination sources. Likewise, it is necessary to ensure the implementation of official controls aiming at the detection of possible contaminants, as well as to define their maximum permissible or at least recommended levels in different categories of food of animal origin, including meat products (Perši et al. 2014). When it comes to dry-cured meat products, quality standardisation and safety of these products call for further research on possible hazard sources, OTA contamination included, especially in view of the fact that the consumption of these nutritious, high-quality traditional meat products is ever growing on a global scale.

Given that many Mediterranean countries have a long tradition of dry-cured meat products' production mostly taking place in rural households, and that these products can be produced under conditions that favour mycotoxin production, further investigations in this field are of an unquestionable interest. Therefore, the aim of this study was to investigate into the occurrence of OTA as the main mycotoxin contaminating different dry-cured meat products, such as dry-fermented sausages and hams produced by many Croatian households.

METHODS

Sampling and Sample Preparation

In general, dry-cured meat products from Croatia include different types of hams (Dalmatian, Istrian, Slavonian, Krk and Drniš ham) and dry-fermented sausages (Istrian, Slavonian and other types of domestic sausages). These products are produced based on traditional recipes and technologies observed by producing rural households located in different parts of Croatia. They are manufactured from the first-, second- and third-category meat without offal, while their production technology includes smoking (except for Istrian sausage and Istrian ham), drying and long-time ripening in darkened chambers (Kovačević 2014; Kovačević 2017). Ripening takes place at the average temperatures of 12-16°C and the relative humidity of 70-80% and lasts 3-4 months in case of dry-fermented sausages or more than 12 months in case of hams.

This study was performed on a total of 135 dry-cured meat product samples, comprising different types of dry-fermented sausages (n = 90) and hams (n = 45) produced by a large number of households in Croatia. The meat products had been randomly collected during 2018 and 2019 from the markets. Each product was sampled in the amount of about 1,000 g, depending on the original product weight. Data on the origin of the raw materials and the production technology were not available. The sampling procedure was carried out in full line with the Commission Regulation (EC 2006b), under which food sampling procedures to the effect of mycotoxin analyses are defined. Immediately after sampling, product surfaces were swabbed for mould identification, followed by OTA analyses. For the sake of OTA analyses, all samples were cut into small pieces and homogenized using a Grindomix GM 200 (Retsch, Haam, Germany). The samples were prepared in line with the ISO 3100-1:1991 (Meat and Meat Products - Sampling and Preparation of Test Samples) and stored at +4 °C prior to OTA analyses.

Mould Isolation and Identification

Visible mycelia of moulds present on the surface of dry-fermented sausages and hams were cut off together with the wrap/surface using a sterile scalpel and transferred into 9-cm Petri dishes containing Dichloran 18%-Glycerol (DG-18). No more than four particles of mouldy wrap/surface were placed onto the DG-18 agar surface. The surface areas lacking any visible colonies were swabbed using sterile damp swabs. To avoid a too crowdie mould growth, damp swabs were dipped into 10 ml Buffered Peptone Water (BPW) with Tween 80, vortexed and transferred into dishes containing DG-18 by spreading/smudging. The inoculated agar media were incubated for five days in darkness at 25 ± 1°C, whereupon individual mycelium colonies were sub-cultivated and transferred into 9-cm (a three-point) Petri dishes containing Dichloran 18%-Glycerol (DG-18) and incubated for another five days in darkness at 25 ± 1°C. In the subsequent course, macroscopic and microscopic characterization of the grown colonies and their genus identification took place. For the purpose of species identification, the identified genera were sub-cultivated on Malt extract agar (MEA) and Czapeak yeast extract agar (CYA) by virtue of three-point inoculation.

Mould isolates were identified at a species level by defining their macroscopic (colour of colony obverse and reverse, diameter, texture, and exudate) and microscopic (the structure of conidiophores, the shape and size of conidia) morphological characteristics. For the determination of micro-morphological characteristics, slides were prepared from the MEA medium using the lactophenol cotton blue (LPCB) as a mounting medium. The slides were analysed using differential interference contrast microscopy under oil immersion at 1,000x magnification. The microscope in use was of an AX10 type (Zeiss, Germany). All isolates were identified according to Pitt and Hocking (2009) and Samson et al. (2004).

Molecular identification of mould isolates was performed in order to verify the results of traditional identification methods. For the DNA extraction purposes, an isolated mould was inoculated into the Sabouraud agar and incubated in darkness at 25 ± 1°C for 7 days. DNA was extracted from about 100 µg of the mould colonies using the DNeasy Plant Mini Kit (Qiagen, Germany) according to the manufacturer's instructions. In molecular identification of mould species, primers specific for ITS, *benA* and *CaM* loci, were selected for PCR amplification (Table 1). Each 25 µL-PCR reaction mix contained 12.5 µL of 2x PCR buffer (HotStarTaq Plus MasterMix Kit, Qiagen, Germany), 2.5 µL of 10x Coral Load, 0.4 µM of each primer, a nuclease-free water and 1 µL of the DNA extract. The procedure took place under the following cycling conditions: 95 °C for 5 min followed by 40 cycles at 94 °C for 30 s, 56 °C for 30 s, 72 °C for 60 s, concluding with the final extension at 72 °C for 10 min. The PCR products were checked using gel electrophoresis in 1.5%-agarose gel stained with GelStar nucleic acid stain (Lonza, Switzerland) and visualized by virtue of

UV trans-illumination. All PCR products of an adequate size were purified prior to sequencing, either using an ExoSAP-IT PCR cleanup reagent (Affymetrix, California, USA) or a QIAquick Gel Extraction Kit (Qiagen, Germany). Purified samples were sent to Macrogen Inc., Amsterdam, the Netherlands, for paired-end sequencing. The obtained sequences were aligned using the Lasergene SeqManPro DNASTAR 13 (Madison, Wisconsin, USA). The edited sequences were compared to those available from the Fungal Biodiversity Centre CBS database and the National Centre for Biotechnology Information (NCBI) GenBank database using the BLAST algorithm.

Table 1. Primers used for mould species identification

Gene	Primer	Sequence(5'→ 3')	Lenght (bp)	Reference
Internal transcribed spacer (ITS)	ITS1	TCCGTAGGTGAACCTGCGG	~600	White et al. 1990
	ITS4	TCCTCCGTCTATTGATATGC		
β-tubulin (*benA*)	Bta_2a	GGTAACCAAATCGGTGCTGCTTTC	~550	Glass and Donaldson 1995
	Bta_2b	ACCCTCAGTGTAGTGACCCTTGGC		
Calmodulin (*CaM*)	CMD5	CCGAGTACAAGGARGCCTTC	~580	Hong et al. 2006
	CMD6	CCGATRGAGGTCATRACGTGG		

Determination of OTA Using the ELISA Method

The ELISA as a quantitative screening method was used in the first stage of OTA analyses of all dry-cured meat products. Sample preparation was performed (with small modifications) as described previously by Lešić et al. (2017). Briefly, into 2.5 g of a sample, 2 mL of 1 M H_3PO_4 and 10 mL of ethyl acetate were added, vigorously shaken and centrifuged (1 min, 2,000 rpm, 20°C). The supernatant was transferred using ethyl acetate (10 mL). After mixing and repeated centrifugation, ethyl acetate layers were combined, supplemented with 10 mL of 0.65 M $NaHCO_3$ and shaken for 30 min. After centrifugation (5 min, 3,000 rpm, 20°C), 4 mL of the aqueous phase were transferred and heated in a water bath (100°C, 5 min) and left to cool. Then, six mL of distilled water were added, and an aliquot portion of the solution was diluted in 3 mL of 0.13 M-$NaHCO_3$.

ELISA kit Ochratoxin A (Art. No. R1311) used for OTA determination was provided by R-Biopharm (Darmstadt, Germany). Each kit contains a micro-titre plate with 96 wells coated with OTA antibodies, OTA aqueous standard solutions (0, 50, 100, 300, 900, and 1,800 ng/L), peroxidase-conjugated OTA, substrate/chromogen solutions, stop-reagents, and dilution/washing buffers. A competitive ELISA test was performed according to the instructions of the kit manufacturer with the use of an auto-analyser (Awareness Technology Inc. 2910, Palm City, USA). After the test, the reaction was stopped by virtue of addition of a stop-reagent (1 N-sulphuric acid) and the absorbance was measured photometrically at 450 nm. Concentrations of OTA (μg/kg), inversely proportional to the absorbance, were calculated from six-point calibration curves plotted by the software the auto-analyser is equipped with, and ultimately corrected for the sample dilution factor and recovery values. All other chemicals and solvents used during the extraction procedures and OTA determination were of an analytical or a HPLC grade.

Validation of the ELISA Method

For the ELISA method, the limit of detection (LOD), the limit of quantification (LOQ), as well as the recovery and the intermediate precision, were determined. OTA standard solutions were prepared from the OTA solid standard (Acros Organics, Geel, Belgium) in form of an aqueous stock and working solution in concentrations of 10,000 μg/L and 20 μg/L, respectively, and stored at +4 °C pending analyses. LOD and LOQ values were calculated from the mean value of concentrations obtained through analyses of six dry-fermented sausages used as control samples (produced by meat industry and first analysed using the LC-MS/MS method) plus two- and ten-fold standard deviation, respectively. The recoveries were determined at three different levels (1.0, 2.0 and 3.0 μg/kg; six replicates per concentration level per day) by virtue of spiking the control samples with the OTA standard working solution (20 μg/L) correspondent to the assessed content levels. As regards the determination of intermediate precision, the same steps were repeated under the same analytical conditions on two additional occasions within a period of three months using different lots of ELISA kits and chemicals.

Determination of OTA Using the LC-MS/MS Method

Samples in which OTA was detected by the ELISA method were further analysed using the LC-MS/MS method. Samples were prepared using highly specific immunoaffinity columns (OCHRAPREP®, R-Biopharm Rhône LTD, Glasgow, Scotland). The mycotoxin extraction procedure was performed according to the instructions of the immunoaffinity columns' manufacturer (OCHRAPREP® - Quantitative detection of Ochratoxin A using HPLC). These immunoaffinity columns contain gel suspension of monoclonal antibodies specific for OTA. After the extraction of the mycotoxin (using a mixture of methanol and 1%-sodium bicarbonate solution (60:40, v/v)) and centrifugation (4,000 rpm, 10 min), meat product samples were filtered and diluted with the phosphate buffer (PBS) containing Tween 20 (7 mL of the filtered sample with 49 mL of 0.01%-Tween 20 PBS) and then applied onto the immunoaffinity columns and passed through them. The next step was column rinsing with (Tween 20-free) PBS (10 mL in case of OTA) aiming to remove the unbound components, followed by the elution of OTA with 1.5 mL of 100%-methanol, back-flushing, and washing with an equal amount of water. Samples were dried under a nitrogen stream at 40°C and reconstituted in 500 μL of $H_2O/ACN/CH_3COOH$ = 49.5/49.5/1.

Preparation of samples for OTA analysis included mycotoxin extraction with the mixture of methanol and 1%-sodium bicarbonate solution (60:40, v/v)), centrifugation (4,000 rpm, 10 min), filtration and dilution 7 mL of the filtered sample with the 49 mL of the phosphate buffer (PBS, pH = 7.4) containing 0.01% - Tween 20. Samples were then applied onto the immunoaffinity columns (OCHRAPREP®, R-Biopharm Rhône LTD,

Glasgow, Scotland) that contain gel suspension of monoclonal antibodies specific for OTA. After passing through columns, samples were rinsed with 10 ml of PBS to remove the unbound components and vacuum-dried (for 4 min). The next step was the elution of OTA with 1.5 mL of 100%-methanol, followed by back-flushing (3x) and washing with 1.5 mL of ultrapure water. Lastly, samples were dried under a nitrogen stream (40°C) and reconstituted in 500 μL of $H_2O/ACN/CH_3COOH$ = 49.5/49.5/1.

The OTA solid standard was obtained from Acros Organics (Geel, Belgium). A high-performance liquid chromatograph (1260 Infinity, Agilent Technologies, Santa Clara, USA) consisting of a degasser, a binary pump, an auto-sampler and a column compartment, was coupled with a triple quadrupole mass spectrometer (6410 QQQ, Agilent Technologies, Santa Clara, USA). Chromatographic separation of OTA was performed on a 150 x 4.6 mm-, 5 μm-particle size Gemini analytical column (Phenomenex, Torrance, USA) coupled with a SecurityGuard™ Cartridges Gemini® C18, 4 x 3.0 mm ID pre-column (Phenomenex, Torrance, USA). The mobile phase A consisted of water and methanol (90/10 (v/v) with 5 mM-ammonium acetate and 1%-acetic acid, while the mobile phase B consisted of methanol and water (97/3 v/v), also with 5 mM-ammonium acetate and 1%-acetic acid. A gradient elution was employed as follows: 0-14 min 100% A, 14-18.10 min 100% B, 18.10-20.5 min 100% A, at the flow rate of 1 mL/min, column temperature of 25°C and the total runtime length of 22 min. The sample injection volume was 40 μL.

A mass spectrometer was operated in the positive ion mode (ESI+), with the source temperature set at 350°C, the gas flow rate set at 12 L/min, the nebulizer set at 20 psi and the capillary voltage set at 2000 V (+) and 4000 V (-). One precursor ion (404) and two product ions (357.9 and 239.0) were monitored. The fragmentor voltage was 130 V and collision energy for each product ion equalled to 25 and 10 eV, respectively. According to the Commission Decision (EC 2002), the requirement imposed on a confirmatory method used for the detection of contaminants is to obtain three identification points. The instrumental method detailed above that monitors two product ions, provides for as much as four identification points.

Validation of the LC-MS/MS Method

The LC-MS/MS method was validated according to the Guidance Document on the Estimation of the LOD and LOQ for measurements in the field of contaminants in feed and food via blank samples (Wenzl et al. 2016). For the LOD/LOQ determination, 10 blank samples of dry-fermented sausages produced by meat industry were used. Samples were spiked with 0.25 µg/kg of OTA, prepared using the above- described procedures and analysed. For each batch, a 5-point calibration curve was plotted. The concentration range of OTA was 0.25, 0.5, 1, 5 and 10 ng/mL. Recovery was determined by virtue of analysing 10 blank samples spiked with OTA at the level of 0.25 µg/kg. The slope of the calibration curve and signal abundances of the spiked samples were then used to calculate LOD and LOQ, while linearity was tested within the concentration ranges stated above. Matrix effect was evaluated by virtue of comparison of sample extract (spiked with 0.5 µg/kg of OTA) responses versus OTA standard responses at same level.

Data Processing

Statistical analysis was performed using the SPSS Statistics Software 22.0 (SPSS Statistics, NY IBM, 2013, Sankt Ingbert, Germany), the statistical significance level thereby being set at 95% ($P = 0.05$).

RESULTS AND DISCUSSION

Validation Study

The LOD and LOQ obtained for the ELISA method were 0.65 µg/kg and 1.10 µg/kg, respectively. The results concerning recovery and intermediate precision are shown in Table 2. The established mean recovery was 108.0%, while the mean intermediate precision equalled to 109.9%. The mean coefficients of variation (CV) established along the line of recovery and intermediate precision determination were 7.9% and 11.6%, respectively. Earlier investigations have shown that the ELISA method can be used as an effective screening method for OTA determination in meat products (Matrella et al. 2006; Markov et al. 2013; Pleadin et al. 2015c). However, an overestimation of OTA concentrations, more pronounced with less spiked samples (1.0 μg/kg) and of significance for the interpretation of the study results, should be kept in mind, too. Therefore, a more substantiated OTA quantification made use of a confirmatory method, employed with all samples revealed by the ELISA assay to be OTA-positive.

Table 2. Recovery and intermediate precision obtained with the validation of the ELISA method for ochratoxin A (OTA) determination

Spiked level (μg/kg)	Recovery (%)	CV (%)	Intermediate precision (%)	CV (%)
1.0	114.3	10.3	116.5	14.1
2.0	108.5	7.1	110.6	10.9
3.0	101.3	6.2	102.5	9.8

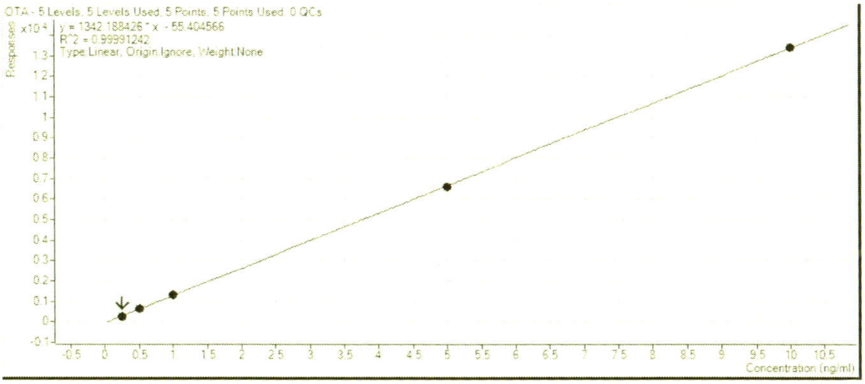

Figure 1. A typical five-point calibration curve for ochratoxin A (OTA) determination.

LC-MS/MS validation resulted in LOD and LOQ of 0.19 μg/kg and 0.63 μg/kg, respectively. The recoveries obtained in this study ranged from 65.37% to 80.74%, with the mean recovery value of 68.39%.

Table 3. Ochratoxin A (OTA) determined in dry-cured meat products[a]

Type of product	Hams (n = 45)	Dry-fermented sausages (n = 90)
No of positives[b]	6	9
Positives (%)	13.3	10.0
Mean (μg/kg)	1.16	0.26
SD (μg/kg)	0.87	0.08
Min (μg/kg)	0.19	0.20
Max (μg/kg)	2.54	0.41

[a] The results represent summary values of both implemented analytical methods: ELISA – for samples in which OTA was not detected; LC-MS/MS for the quantitative determination in samples in which OTA was first detected using the ELISA method.

[b] Positives - samples in which OTA concentration ascertained using the LC-MS/MS method was determined to be >LOD (0.19 μg/kg).

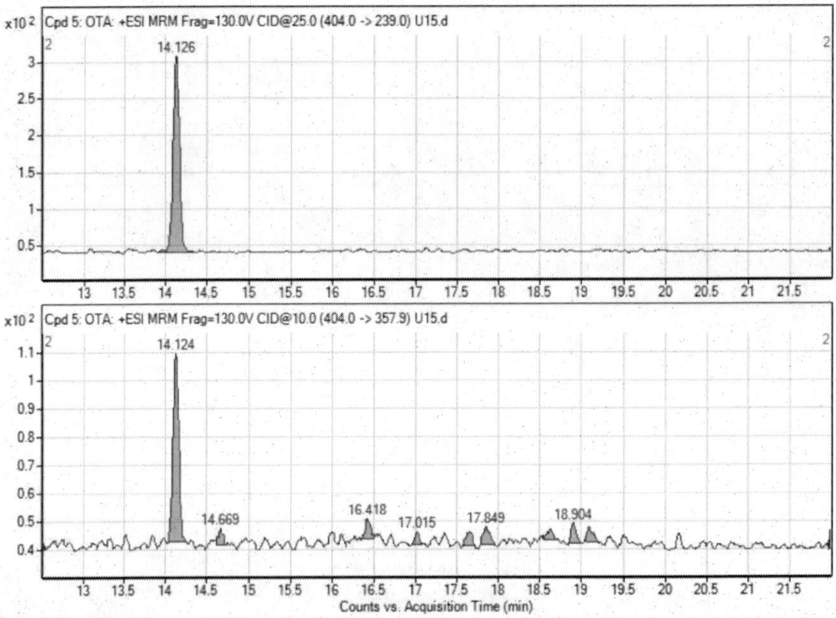

Figure 2. MRM chromatograms of OTA in a sample spiked with 0.5 μg/kg of OTA.

Matrix effect evaluation revealed 17.8% ion enhancement, so that the results obtained during the recovery determination were corrected accordingly. The results met the requirement for method trueness defined for the recovery of ≤ 1 μg/kg (-50% to + 20%) (EC 2002). An adequate optimization of diagnostic ions conformant to the mass spectrometer settings, as well as an adequate sample preparation technique, resulted in high-sensitive chromatograms. Figure 1 shows the OTA (0.5 μg/kg) MRMs in a sample spiked for matrix effect evaluation. Figure 2 shows a typical five-point calibration curve for OTA determination. The regression coefficient (R^2) was ≥0.999, showing an adequate linearity in 0.25 to 10 ng/mL calibration range. On top of adequate linearity, the slope of the calibration curve also shows an appropriate measurement sensitivity.

Moulds Identified on the Surfaces of TMPs

From the surface of different dry-cured meat products comprised by this study, moulds of the *Penicillium,* the *Aspergillus* and the *Mucor* genus were isolated. A significantly higher ($p < 0.05$) number of total isolates was identified in hams (69%) in comparison with dry-fermented sausages (31%) (Figure 3), which can be explained by a considerably longer ham ripening period (of at least 12 months and often up to 18 months) as compared to that of dry-fermented sausages (2-4 months), which gives different moulds a better chance to overgrow ham surfaces. On top of that, ham production goes on all over the four seasons, while sausage production takes place exclusively in winter months. The contribution of OTA-producers to total isolated mould species is shown in Figure 4.

Earlier studies performed in other European countries have evidenced that during the ripening period surfaces of dry-cured meat products are typically overgrown with moulds that mainly belong to the *Penicillium* and the *Aspergillus* genera, and that the *Aspergillus* spp. are less frequently isolated than the *Penicillium* spp. These moulds are known for their occasional uncontrolled growth that may facilitate the development of an off-flavour and cause a direct mycotoxin contamination of the meat product (Comi et al. 2004; Dall'Asta et al. 2010; Markov et al. 2013; Rodríguez et al. 2012b; Sonjak et al. 2011; Tabuc et al. 2004).

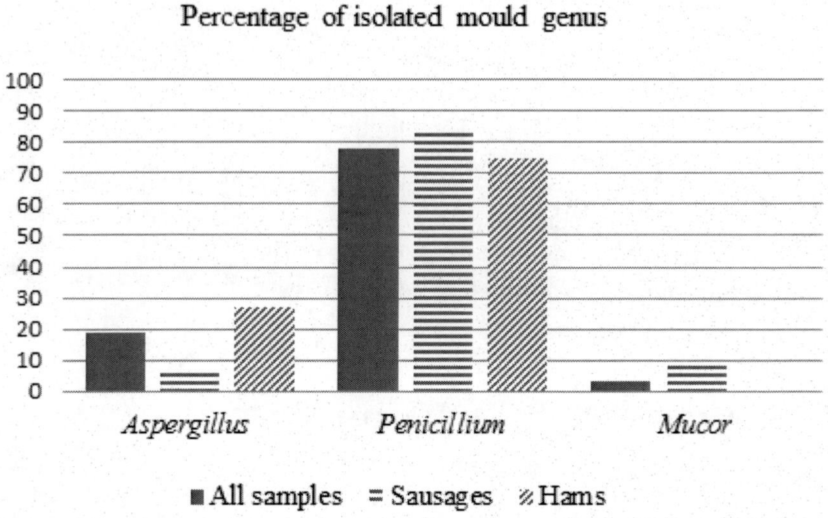

Figure 3. The percentage of mould genera isolated from the surface of the studied dry-fermented sausages and hams.

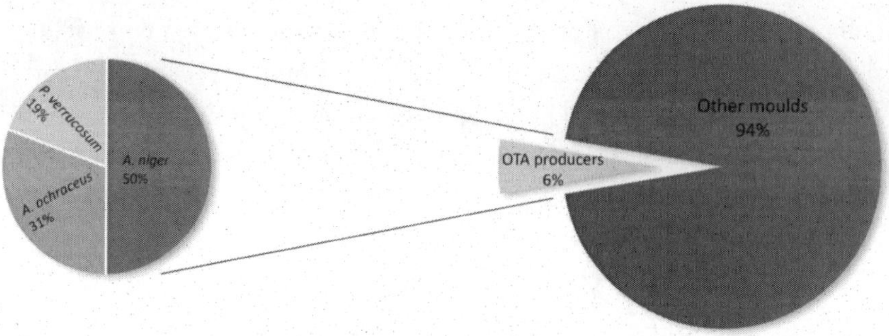

Figure 4. Contribution of ochratoxin A (OTA)- producers to total isolated mould species.

The number of isolates identified in this study shows the domination of the *Penicillium* genus (78%), whereas species of other genera were present in significantly lower percentages, as follows: *Aspergillus* (19%) and *Mucor* (3%). When results are displayed based on the product type, *Penicillium* species dominate in both ham and dry-fermented sausage samples. However, the representation of other mould species in meat products under consideration varies as compared to totals established for the two combined.

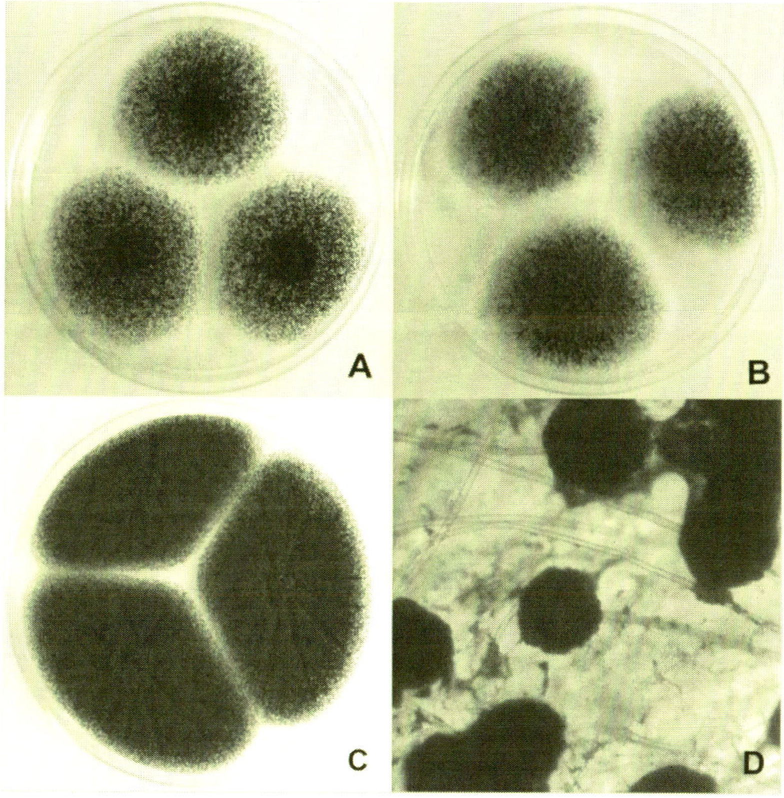

Figure 5. Ochratoxin A (OTA)-producer *Aspergillus niger* on DG 18 agar (A), MEA (B), CYA (C), micrograph, magnification 100x (D).

In ham samples, *Aspergillus* species are isolated in a higher percentage (17%) than in fermented sausage samples (2%). At the same time, in fermented sausages *Mucor* species were isolated in 15 samples (3%), as oppose to ham samples that were unanimously *Mucor*-free. That difference could be explained by the difference in length and season of ripening of the studied products. It is known that ham production goes on all over the four seasons, including summertime characterised by high temperatures and dry weather preferable to the *Aspergillus* species (Pitt and Hocking 2009). Therefore, the presence of *A. niger* (50%) (Figure 5) and *A. ochraceus* (31%) (Figure 6) as the most frequently encountered OTA- producers was to be expected.

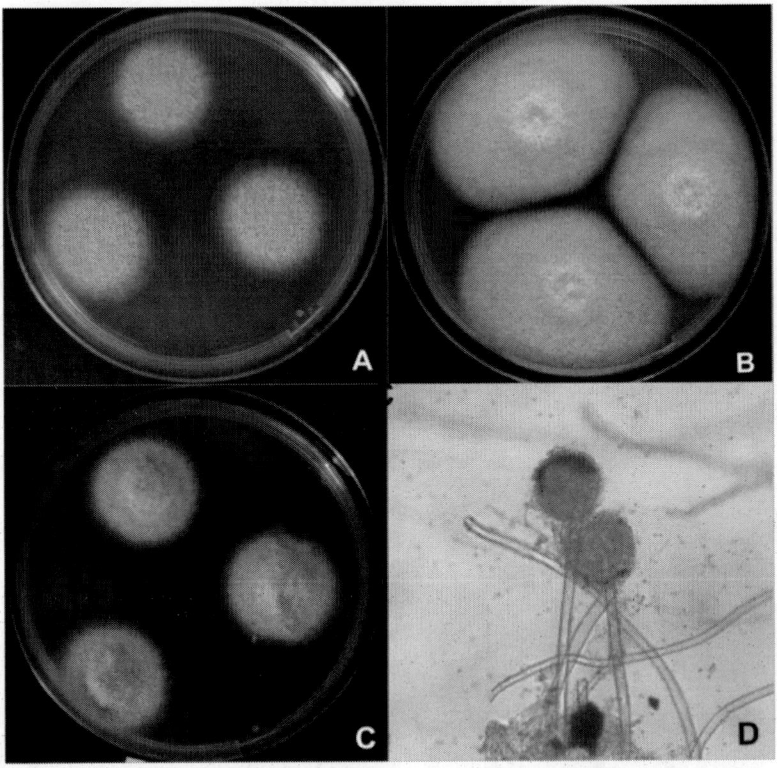

Figure 6. Ochratoxin A (OTA)-producer *Aspergillus ochraceus* on DG 18 agar (A), MEA (B), CYA (C), micrograph, magnification 100x (D).

In dry-fermented sausages, the only potential OTA-producer present was *A. niger*. Therefore, it could be concluded that *A. niger* produced OTA in a colder environment in which the production of dry-fermented sausages takes place, while *P. verrucosum* (19%-isolation) (Figure 7) started to produce OTA in warmer environments in which hams are produced. Explanation for the growth of the *Mucor* species on fermented sausages can be found in the fact that the *Mucor* species can grow in an atmosphere having a low O_2 content and ≥97% of CO_2 (Pitt and Hocking 2009). These particular conditions are met during the smoking stage of the sausage producing. Also, *Mucor* species growth is enhanced by high moisture, i.e., water activity (a_w) on a sausage surface higher than 0.93 (Morin-Sadrin et al. 2016), contrary to *Aspergillus*, which prefers the a_w to be higher than 0.80 (Ribeiro et al. 2006).

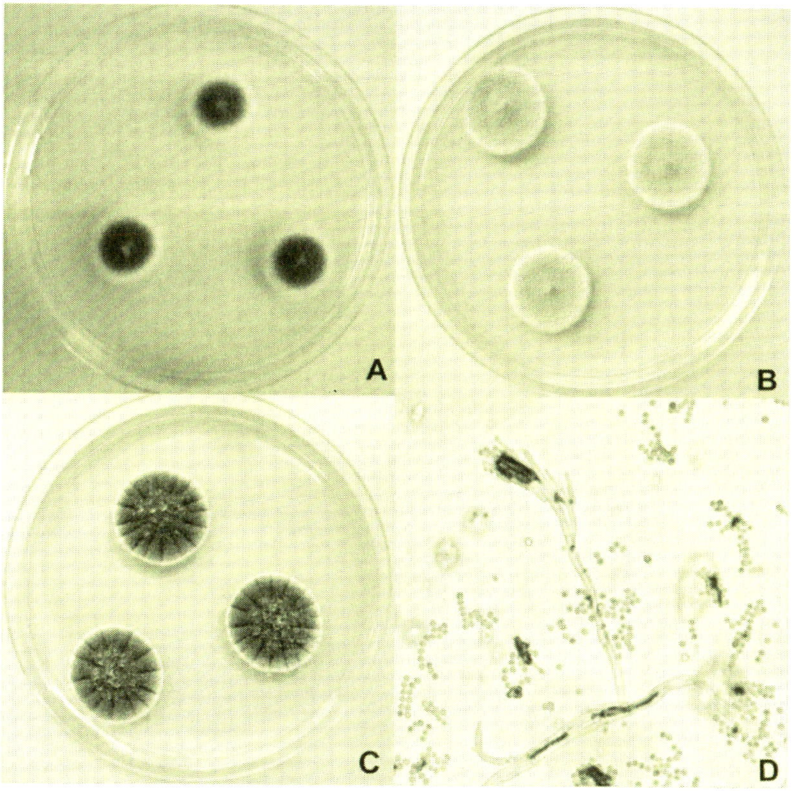

Figure 7. Ochratoxin A (OTA)-producer *Penicillium verrucosum* on DG 18 agar (A), MEA (B), CYA (C), micrograph, magnification 100x (D).

OTA Contamination

OTA concentrations determined in dry-cured meats of this study are presented in Table 2. Figure 8 shows OTA MRM chromatograms of a) a ham sample contaminated with OTA at the level of 2.54 μg/kg and b) a blank ham sample. Figure 9 shows chromatograms obtained with a) a dry-fermented sausage sample contaminated with OTA at the level of 0.41 and

b) a blank dry-fermented sausage sample. These concentrations represent the maximal obtained OTA levels for each of the two analysed types of dry-cured meat products.

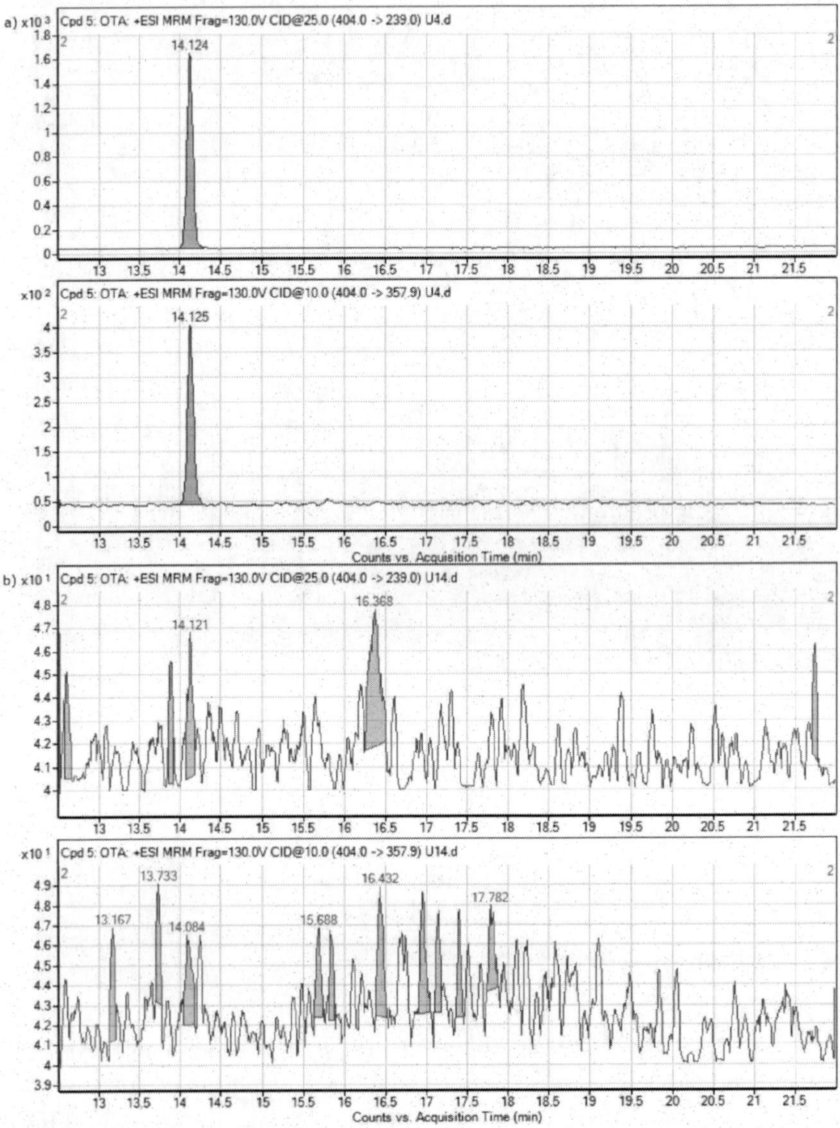

Figure 8. MRM chromatograms for OTA in a) a ham sample contaminated with OTA (2.54 μg/kg) and b) a blank ham sample.

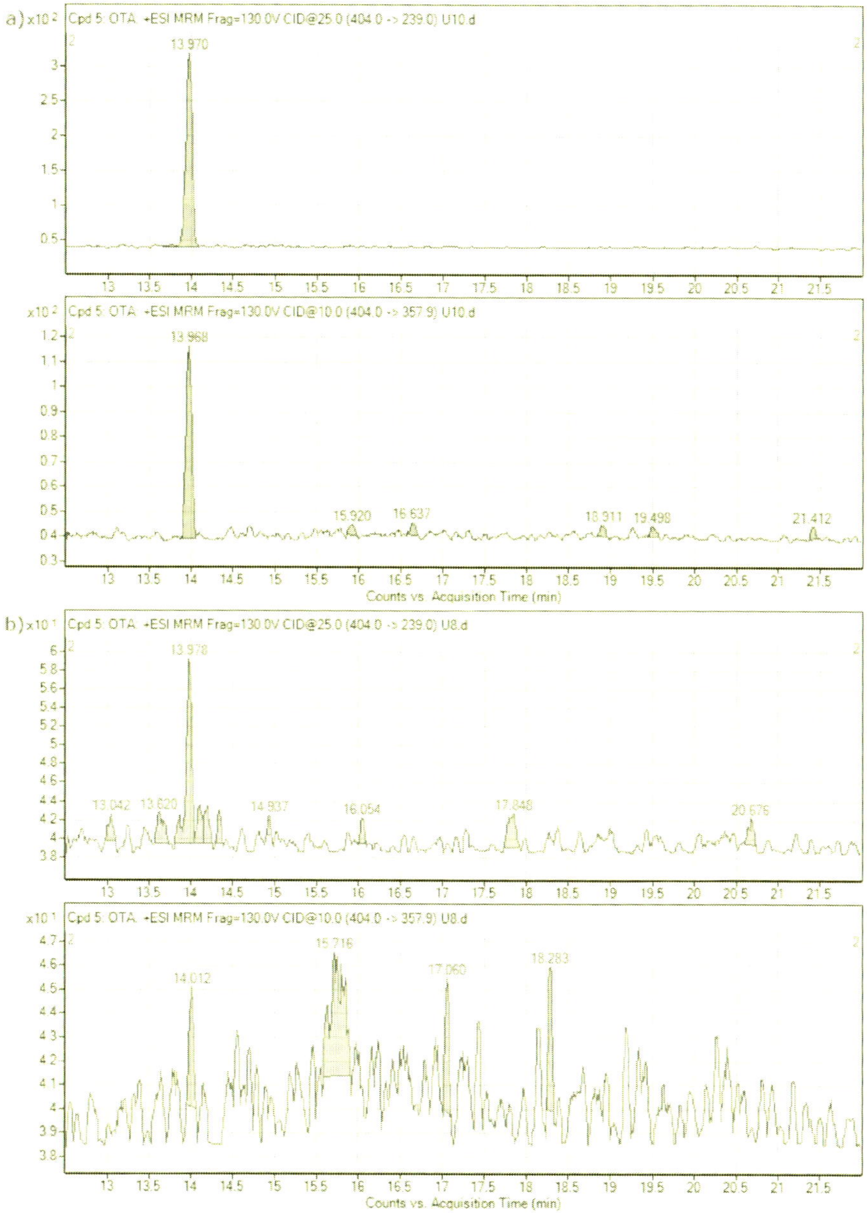

Figure 9. MRM chromatograms for OTA in a) a sausage sample contaminated with OTA (0.41 μg/kg) and b) a blank sausage sample.

OTA concentrations obtained in dry-fermented sausages ranged from 0.20 μg/kg to the maximal of 0.41 μg/kg, while in the analysed hams they spanned from 0.19 μg/kg to 2.54 μg/kg. In total, 11% of all traditional meat products were positive on OTA (>LOD). Besides a possible primary contamination of the products consequential to the presence of toxicogenic moulds evidenced in four samples (3 hams and 1 fermented sausage), we can assume that other products on whose surfaces toxicogenic moulds were not identified, were probably contaminated with OTA in form of secondary contamination (carry-over effect) that can come as a result of contaminated feed the animals were fed on. However, as of now no Maximum Levels (ML) for mycotoxins in meat and meat products have been stipulated by regulatory bodies (EC 2006a). However, within the frame of their national legislation, some of the EU member states, for instance Italy and Denmark, have laid down the maximal recommendable level of 1 μg/kg of mycotoxins, tolerable to be present in this kind of food (that is to say, meat, meat products and offal). In this study, concentrations higher than 1 μg/kg were obtained in three ham samples, but in none of the sausage samples.

Earlier studies performed in European countries have shown that mycotoxin contamination of raw pork and final pork products is quite often, so that it should be monitored (Gareis and Scheuer 2000; Chiavaro et al. 2002; Pietri et al. 2006; Pleadin et al. 2015c). While mycotoxins in meat primarily occur as a result of indirect transmission from naturally contaminated feed, in case of meat products their occurrence is often influenced by the recipe used in their production, i.e., the origin of meat and the use of edible tissues, blood and spices which can be contaminated with mycotoxins (Gareis and Scheuer 2000), as well as by the production and storage conditions (Asefa et al. 2011; Rodríguez et al. 2012b).

Our previous research showed that the presence of OTA in meat products may arise as a consequence of farm animal feeding on contaminated feed or the use of spices during the production process (Pleadin et al. 2013; Perši et al. 2014). Following an OTA treatment of fattening pigs, a transfer from OTA-contaminated feed into first- and second-category meat and final meat products was evidenced. Our research on OTA presence in fermented sausages using the ELISA method resulted with concentrations to 6.68 μg/kg found in Slavonian sausages (Markov et al. 2013; Vulić et al. 2014) and 5.17 μg/kg found in Slavonian Kulen (Frece et al. 2010; Vulić et al. 2014; Pleadin et al. 2016). The research on mycotoxin prevalence conducted on a larger number of various dry and semi-dry traditional meat products sampled within the 2011-2014 timeframe, showed the OTA span of 1.23 μg/kg in sausages to 9.95 μg/kg in prosciutto (Pleadin et al. 2015c). Our last research devoted to the prevalence of OTA in meat products, established that during a long ripening of the dry-fermented sausage called Slavonski Kulen, the nascence of mycotoxins may come as a result of an unwelcome presence of superficial mould species, which, in certain (uncontrolled) environments produce these mycotoxins during several months of ripening (Pleadin et al. 2017). However, in the studies quoted above the occurrence of OTA was not put into direct connection with the evidenced presence of toxicogenic moulds, which may be responsible for the final dry-cured meat product contamination.

The studies referred to above came up with the conclusion that meat products' contamination with OTA arises on the grounds of inadequate production control and inadequate storage practices, imposing the need for prevention, systematic control and further monitoring, so as to be able to pinpoint the circumstances facilitating the production of this mycotoxin. Research conducted in other European countries on certain types of dry-cured meat products (Núñez et al. 1996; Asefa et al. 2009; Sonjak et al. 2011; Iacumin et al. 2011) has tackled the issue of superficial mould presence and mycotoxin contamination only partly, neglecting thereby to discuss it in view of climate-conditioned mould prevalence in targeted geographical regions. Furthermore, it has been established that outer casing damages also facilitate OTA diffusion into the dry-fermented sausage interior (Pleadin et al. 2015a; Pleadin et al. 2015b).

In order to prevent the occurrence of mycotoxins in dry-cured meat products, the growth of toxicogenic moulds should be well-controlled (Núñez et al. 2015; Pleadin et al. 2015c). Moulds colonising the product surface should be continuously removed from it by virtue of brushing and rinsing during the entire ripening period, so as to prevent an excessive mouldiness of the product surface (Sørensen et al. 2008). Given that consumers usually prefer surface mould-free products, producers tent to rinse semi-cured meat products at some point between the drying and the ripening stage, so as to remove visible mould colonies inhabiting the product surface (Asefa et al. 2011). OTA concentration established on an outer surface (casing) of regularly brushed and rinsed dry-cured sausages was found to be below the limit of detection of an analytical method employed. One of the common practices is to spray rice flour over the surface of ripened sausages upon mould removal via brushing, rinsing or cleansing under pressurised air. Iacumin et al. (2009) are of the opinion that brushing of dry-cured sausage surfaces should come first and then be followed by their rinsing, so as to reduce OTA concentration and eliminate any potential threat to consumer health. On top of that, in order to prevent excessive mouldiness of the product surface during the ripening stage, it is necessary to secure the distance between the products that shall render their mutual contact impossible and let the air flow in an unhindered manner. Products should be ripened in ripening chambers equipped with microfilters that allow the inflow of fresh air. Ripening chamber surfaces should be coated with fungicidal coats (prior to products' delivery and ripening commencement), while the chamber entrance point should be secured by a pressure barrier that prevents the influx of an outer air.

Conclusion

Production conditions seen in rural households producing traditional meat products are mostly uncontrolled, which allows for a significant influence of external factors on the prevalence of surface moulds, in particular toxicogenic moulds whose presence can result in mycotoxin contamination of final products. This study managed to prove the presence of OTA in final dry-cured meat products, coming as a consequence of the presence of toxicogenic moulds on the surface of these products. Other factors of influence are the duration of ripening and climatic conditions of the production region. In order to prevent possible harmful impact on consumer health, during the production and storage of meat products, in particular during their ripening, all conceivable preventative activities should be taken to prevent mycotoxin contamination and to control mould occurrence. Further research should deal with factors influencing OTA presence, in particular with uncontrolled production environments seen in rural households. On top of OTA, future research should comprise other, less investigated mycotoxins such as citrinin, sterigmatocystin and cyclopiazonic acid, which, according to literature sources, can be present in meat products in high concentrations.

ACKNOWLEDGMENT

This work was funded by the Croatian Science Foundation through the project "Mycotoxins in traditional Croatian meat products: molecular identification of mycotoxin-producing moulds and consumer exposure assessment" (HRZZ-IP-2018-01-9017).

REFERENCES

Amézqueta, Susana, Elena Gonzàles-Peñas, Marìa Murillo-Arbizu, and Adela L. de Cerain. 2009." Ochratoxin A decontamination: A review." *Food Control* 20: 326-333.

Andersen, Susanne J. 1995. "Compositional changes in surface mycoflora during ripening of naturally fermented sausages." *Journal of Food Protection* 58: 426–429.

Asefa, Dereje T., Cathrine F. Kure, Ragnhild O. Gjerde, Mohamed K. Omer, Solveig Langsrud, Truls Nesbakken, and Ida Skaar. 2010. "Fungal growth pattern, sources and factors of mould contamination in a dry-cured meat production facility." *International Journal of Food Microbiology* 140: 131-135.

Asefa, Dereje T., Cathrine F. Kure, Ragnhild O. Gjerde, Solveig Langsrud, Mohamed K. Omer, Truls Nesbakken, and Ida Skaar. 2011. "A HACCP plan for mycotoxigenic hazards associated with dry-cured meat production processes." *Food Control* 22: 831-837.

Asefa, Dereje T., Ragnhild O. Gjerde, Maan S. Sidhu, Solveig Langsrud, Cathrine F. Kure, Truls Nesbakken, and Ida Skaar. 2009. "Moulds contaminants on Norwegian dry – cured meat products." *International Journal of Food Microbiology* 128*:* 435 – 439.

Battilani, Paola, Amedeo Pietri, Paola Giorni, Silvia Formenti, Terenzio Bertuzzi, Tania Toscani, Roberta Virgili, and Zofia Kozakiewicz. 2007. "*Penicillium* populations in dry-cured ham manufacturing plants." *Journal of Food Protection* 7: 975-980.

Bertuzzi, Terenzio, Gualla, Alessia, Mauro Morlacchini, and Amedeo Pietri. 2013. "Direct and indirect contamination with ochratoxin A of ripened pork products." *Food Control* 34: 79-83.

Bruna, Josè M., Eva M. Hierro, Lorenzo de la Hoz, Donald S. Mottram, Manuela Fernández, and Juan A. Ordóñez. 2003. "Changes in selected biochemical and sensory parameters as affected by the superficial inoculation of *Penicillium camemberti* on dry fermented sausages." *International Journal of Food Microbiology* 85: 111-125.

Bullerman, Lloyd, and Andreia Bianchini. 2007. "Stability of mycotoxins during food processing." *International Journal of Food Microbiology* 119: 140-146.

Chiavaro, Emma, A. Lepiani, F. Colla, P. Bettoni, E. Pari, and E. Spotti. 2002. "Ochratoxin A determination in ham by immunoaffinity clean-up and a quick fluorometric method." *Food Additives and Contaminants* 19: 575-581.

Comi, Giuseppe, and Lucilla Iacumin. 2013. "Ecology of moulds during the pre-ripening and ripening of San Daniele dry cured ham." *Food Research International* 54: 1113-1119.

Comi, Giuseppe, Sandi Orlić, Sulejman Redžepović, Rosalinda Urso, and Lucilla Iacumin. 2004. "Moulds isolated from Istrian dried ham at the pre-ripening and ripening level." *International Journal of Food Microbiology* 96*:* 29-34.

Dall'Asta, Chiara, Gianni Galaverna, Terenzio Bertuzzi, Alessandra Moseriti, Amedeo Pietri, Arnaldo Dossena, and Rosangela Marchelli.

2010. "Occurrence of ochratoxin A in raw ham muscle, salami and dry-cured ham from pigs fed with contaminated diet." *Food Chemistry* 120: 978-983.

Duarte, Sofia C., Angelina Pena, and Celeste M. Lino. 2010. "Ochratoxin A in Portugal: A review to assess human exposure." *Toxins* 2:1225–1249.

European Commission (EC). 2002. "Commission Decision No 2002/657 of 12 August 2002 implementing Council Directive 96/23/EC concerning the performance of analytical methods and the interpretation of results. "*Official Journal of the European Union* L 221/8-36.

European Commission (EC). 2006a. "Commission Regulation No 1881/2006 of 19 December 2006 setting maximum levels for certain contaminants in foodstuffs." *Official Journal of the European Union* L 364/5-26.

European Commission (EC). 2006b. "Commission Regulation No 401/2006 of 23 February 2006 laying down the methods of sampling and analysis for the official control of the levels of mycotoxins in foodstuffs." *Official Journal of the European Union* L 70/12-34.

Frece, Jadranka, Ksenija Markov, and Dragan Kovačević. 2010. "Determination of indigenous microbial populations, mycotoxins and characterization of potential starter cultures in Slavonian kulen." *Meso* 12: 92-99.

Gareis, M., and R. Scheuer. 2000. "Ochratoxin A in meat and meat products." *Archiv für Lebensmittelhygien* 51: 102-104.

Glass, Louise N., and Gary C. Donaldson. 1995. "Development of primer sets designed for use with the PCR to amplify conserved genes from filamentous ascomycetes." *Applied Environmental Microbiology* 61: 1323–1330.

Hong, Seung B., Hye-Sun Cho, Hyeon-Dong Shin, Jens C. Frisvad, Robert A. Samson. 2006. "Novel Neosartorya species isolated from soil in korea." *International Journal of Systematic and Evolutionary Microbiology* 56: 477–486

Hussein, Hussein S., and Jefrey M. Brasel. 2001. "Toxicity, metabolism, and impact of mycotoxins on humans and animals." *Toxicology* 167: 101-134.

Iacumin, Lucilla, Luca Chiesa, Daria Boscolo, Marisa Manzano, Carlo Cantoni, Sandi Orlić, and Giuseppe Comi. 2009. "Moulds and ochratoxin A on surfaces of artisanal and industrial dry sausages." *Food Microbiology* 26: 65-70.

Iacumin, Lucilla, Serena Milesi, Silvia Pirani, Giuseppe Comi, and Luca M. Chiesa. 2011. "Ochratoxigenic mold and ochratoxin an in fermented sausages from different areas in Northern Italy: Occurrence, reduction or prevention with ozonated air." *Journal of Food Safety* 31:538-545.

International Agency for Research on Cancer (IARC). 1993. *Some Naturally Occurring Substances: Food Items and Constituents, Heterocyclic Aromatic Amines and Mycotoxins. IARC Monographs on the Evaluation of Carcinogenic Risks to humans, Vol. 56.* Lyon: IARC Press.

International organization for standardization (ISO). 1991. ISO 3100-1:1991. Meat and meat products - Sampling and preparation of test samples.

Iqbal, Shahzad Z., Sonia Nisar, Muhammad R. Asi, and Selamat Jinap. 2014. "Natural incidence of aflatoxins, ochratoxin A and zearalenone in chicken meat and eggs." *Food Control* 43: 98-103.

Joint FAO/WHO Expert Committee on Food Additives (JECFA). 2001. *Safety evaluation of certain mycotoxins in food. Fifty-six report*. WHO Technical Report Series, Geneva: World Health Organization.

Kovačević, Dragan. 2014. *Technology of kulen and other fermented sausages.* Osijek: Faculty of Food Technology.

Kovačević, Dragan. 2017. *Chemistry and technology of hams and prosciutos.* Osijek: Faculty of Food Technology.

Lešić, Tina, Greta Krešić, Sanja K. Kravar, and Jelka Pleadin. 2017. Nutritional quality of fat in industrial sausages. *Meso* 19 (6): 496-503.

Marin, Sonia, A. J. Ramos, German Cano-Sancho, and Vicente Sanchis. 2013. Mycotoxins: Occurrence, toxicology, and exposure assessment. *Food and Chemical Toxicology* 60:218-237.

Markov, Ksenija, Jelka Pleadin, Martina Bevardi, Nada Vahčić, Darja Sokolić-Mihalek, and Jadranka Frece. 2013. "Natural occurrence of aflatoxin B1, ochratoxin A and citrinin in Croatian fermented meat products." *Food Control* 34:312-317.

Matrella, R., Linda Monaci, M. A. Francesco Palmisano, and Giuseppina M. Tantillo. 2006. "*Ochratoxin A* determination in paired kidneys and muscle samples from swines slaughtered in southern Italy." *Food Control* 17 (2): 114-117.

Morin-Sardin, Stèphanie, Karim Rigalma, Louis Coroller, Jean-Luc Jany, and Emmanuel Coton. 2016. "Effect of the temperature, pH, and water activity on Mucor spp. Growth on synthetic medium, cheese analog and cheese analog and cheese." *Food Microbiology* 56: 69-79.

National Centre for Biotechnology Information (NCBI) GenBank database. Accessed September 20, 2019. http://blast.ncbi.nlm.nih.gov/Blast.cgi.

Núñez, Fèlix, Mar M. Rodríguez, Elena M. Bermúdez, Juan J. Córdoba, and Miguel A. Asensio. 1996. "Composition and toxigenic potential of the mould population on dry-cured Iberian ham." *International Jornal of Food Microbiology* 32: 185–197.

Núñez, Fèlix, Marìa S. Lara, Belèn Peromingo, Josuè Delgado, Lourdes Sanchez-Montero, and Marìa J. Andrade. 2015. "Selection and evaluation of *Debaryomyces hansenii* isolates as potential bioprotective agents against toxigenic penicillia in dry fermented sausages." *Food Microbiology* 46: 114-120.

Ockerman, Herbert W., F. J. Céspedes Sánchez, and F. León Crespo. 2000. "Influence of molds on flavor quality of Spanish ham." *Journal of Muscle Foods* 11: 247-259.

Perši, Nina, Jelka Pleadin, Dragan Kovačević, Giampiero Scortichini, and Salvatore Milone. 2014." Ochratoxin A in raw materials and cooked meat products made from OTA-treated pigs." *Meat science* 96:203-210.

Pietri, Amedeo, Terenzio Bertuzzi, Alessia Gualla, and Gianfranco Piva. 2006. "Occurrence of ochratoxin a in raw ham muscles and in pork products from Northern Italy." *Italian Journal of Food Science* 18: 99–106.

Pitt, John I., and Ailsa D. Hocking. 2009. *Fungi and Food Spoilage.* (3rd ed.). New York: Springer.

Pleadin, Jelka, Nina Perši, Dragan Kovačević, Nada Vahčić, Giampiero Scortichini, and Salvatore Milone. 2013. "Ochratoxin A in traditional dry-cured meat products produced from subchronic-exposed pigs." *Food Additives and Contaminants: Part A* 30:1827-1836.

Pleadin, Jelka, Nina Perši, Dragan Kovačević, Ana Vulić, Jadranka Frece, and Ksenija Markov. 2014. "Ochratoxin A reduction in meat sausages using processing methods practiced in households." *Food Additives and Contaminants: Part B* 7: 239-246.

Pleadin, Jelka, Dragan Kovačević, and Nina Perši. 2015a. "Ochratoxin A contamination of the autochthonous dry-cured meat product "Slavonski Kulen" during a six-month production process." *Food Control* 57:377-384.

Pleadin, Jelka, Dragan Kovačević, and Irena Perković. 2015b. "Impact of casing damaging on aflatoxin B1 concentration during the ripening of dry-fermented sausages." *Journal of Immunoassay and Immunochemistry* 36:655-666.

Pleadin, Jelka, Mladenka Malenica Staver, Nada Vahčić, Dragan Kovačević, Salvatore Milone, Lara Saftić, and Giampiero Scortichini. 2015c. "Survey of aflatoxin B1 and ochratoxin A occurrence in traditional meat products coming from Croatian households and markets." *Food Control* 52: 71-77.

Pleadin, Jelka, Manuela Zadravec, Mario Mitak, Irena Perković, Mario Škrivanko, and Dragan Kovačević. 2016. "Mould contamination of the Croatian autochthonous dry-cured meat product "Slavonski Kulen"." *Book of Proceedings of the 18th World Congress of Food Science and Technology IUFoST 2016*: 346. Dublin: Institute of Food Science and Technology of Ireland.

Pleadin, Jelka, Manuela Zadravec, Dragan Brnić, Irena Perković, Mario Škrivanko, and Dragan Kovačević. 2017. "Moulds and mycotoxins detected in the regional speciality fermented sausage "slavonski kulen" during a 1-year production period." *Food Additives and Contaminants Part A* 34: 282-290.

Ribeiro, J.M., Cavaglieri, L.R., Fraga, M.E., Direito, G.M., Dalcero, A.M., Rosa, C.A. 2006. "Influence of water activity, temperature and time on mycotoxins production on barley rootlets." *Letters in Applied Microbiology* 42: 179-184.

Richard, John L. 2007. "Some major mycotoxins and their mycotoxicoses – An overview." *International Journal of Food Microbiology* 119: 3-10.

Rodríguez, Alicia, Mar Rodríguez, Alberto Martín, Fèlix Nuñez, and Juan J. Córdoba. 2012a. "Evaluation of hazard of aflatoxin B1, ochratoxin A and patulin production in dry-cured ham and early detection of producing moulds by qPCR." *Food Control* 27: 118-126.

Rodríguez, Alicia, Mar Rodríguez, Alberto Martín, Josuè Delgado, and Juan J. Córdoba. 2012b. "Presence of ochratoxin A on the surface of dry-cured Iberian ham after initial fungal growth in the drying stage." *Meat Science* 92: 728-734.

Rodríguez, Alicia, Daniela Capela, Ángel Medina, Juan J. Córdoba, and Naresh Magan. 2015. "Relationship between ecophysiological factors, growth and ochratoxin A contamination of dry-cured sausage based matrices." *Internal Journal of Food Microbiology* 194: 71-77.

Samson, Robert A., E. S. Hoekstra, and Jens C. Frisvad. (2004). *Introduction to food- and airborne fungi*. (7th ed.). Utrecht: CBS.

Sonjak, Silva, Mia Ličen, Jens C. Frisvad, and Nina Gunde-Cimerman. 2011. "The mycrobiota of three dry-cured meat products from Slovenia." *Food Microbiology* 28: 373-376.

Sørensen, Louise M., Tomas Jacobsen, Per V. Nielsen, Jens C. Frisvad, and Anette G. Koch. 2008. "Mycobiota in the processing areas of two different meat products." *International Journal of Food Microbiology* 124: 58-64.

Tabuc, C., Jean D. Bailly, S. Bailly, A. Querin, and P. Guerre. 2004. "Toxigenic potential of fungal mycoflora isolated from dry cured meat products: Preliminary study." *Revue de Médicine Véterinaire* 156: 287-291.

Toldrá, Fidel. 1998. "Proteolysis and lipolysis in flavour development of dry cured meat products." *Meat Science* 49: 101-110.

Vulić, Ana, Nina Perši, Nada Vahčić, Brigita Hengl, Andreja Gross-Bošković, A., Jurković, Dragan Kovačević, and Jelka Pleadin. 2014. "Assessment of possible exposure to ochratoxin A through consumption of contaminated meat products." *Meso* 16:138-144.

Wenzl, Thomas, Johannes Haedrich, Alexander Schaechtele, Piotr Robouch, and Joerg Stroka. 2016. Guidance document on the estimation of LOD and LOQ for measurements in the filed of contaminants in feed and food. EUR 28099. *Publications Office of the European Union.* doi:10.2787/8931.

Westerdijk Fungal Biodiversity. CBS database. Accessed September 20, 2019. http://www.cbs.knaw.nl.

White, Thomas J., Thomas Bruns, Steven B. Lee, and John W. Taylor. 1990. "Amplification and direct sequencing of fungal ribosomal RNA genes for phylogenetics. In *PCR protocols: a guide to methods and applications,* edited by Michael A. Innis, Davis H. Gelfand, John J. Sninsky, and Thomas J. White, 315–322. New York: Academic Press Inc.

Biographical Sketch

Jelka Pleadin, PhD

Affiliation: Croatian Veterinary Institute, Zagreb

Business Address: Savska Cesta 143, 10000 Zagreb

Personal Information:

Surname and Name: Pleadin Jelka
Researchers Identification Number: 273891
URL of web site: https://bib.irb.hr/lista-radova?autor=273891

Education:

2006: PhD degree in Biotechnical Sciences, field of science: Food Technology Academic institution: Faculty of Food Technology and Biotechnology University of Zagreb.

1998: Master of Biotechnology Academic institution: Faculty of Food Technology and Biotechnology University of Zagreb.

Employment:

From 2016 on: Scientific Adviser, Associate Professor (substantive position). Institution: Croatian Veterinary Institute, Zagreb.

2011 – 2016: Scientific Adviser, Assistant Professor (substantive position). Institution: Croatian Veterinary Institute, Zagreb.

2010 – 2011: Senior Scientific Associate. Institution: Croatian Veterinary Institute, Zagreb.

2007 – 2010: Scientific Associate. Institution: Croatian Veterinary Institute, Zagreb.

2006 – 2007: Senior Assistant. Institution: Croatian Veterinary Institute, Zagreb.

2004 – 2006: Assistant. Institution: Croatian Veterinary Institute, Zagreb.

2001 – 2004: Junior Assistant. Institution: Croatian Veterinary Institute, Zagreb.

2001 – 2007: Junior Researcher. Institution: Croatian Veterinary Institute, Zagreb.

Fellowships and Awards:

2016: State Science Award in the field of biotechnical sciences for an exceptionally rich scientific, teaching and professional activities in 2016.

Supervision of Graduate/Doctoral Students and Postdoctoral Researchers:

2010 – 2019: 3 graduate students, 3 doctoral students, 1 Rector's Award-winning contribution. Academic institution: Faculty of Food Technology and Biotechnology University of Zagreb, Croatian Veterinary Institute Zagreb.

2011- 2017: 2 post-doctoral students. Institution: Croatian Veterinary Institute Zagreb.

2012– 2017: 1 graduate student, 2 doctoral students. Academic institution: Faculty of Food Technology "J J Strossmayer" University of Osijek, Croatian Veterinary Institute Zagreb.

2015 – 2017: 4 graduate students. Academic institution: Faculty of Veterinary Medicine University of Zagreb, Croatian Veterinary Institute Zagreb.

2015 – 2017: 4 graduate students. Institution: Department of Biotechnology University of Rijeka, Croatian Veterinary Institute Zagreb.

2015 – 2017: 3 graduate students. Academic institution: School of Medicine University of Rijeka, Croatian Veterinary Institute Zagreb.

2016 – 2017: 4 graduate students. Academic institution: Faculty of Agronomy and Food Technology University of Mostar.

Teaching Activities:

From 2017 on: Lecturer delivering tuition within the frame of postgraduate PhD studies in Food Technology & Nutrition Science hosted by the Faculty of Food Technology "J J Strossmayer" University of Osijek; Course: Instrumental techniques employed in food analytics.

From 2015 on: Course Coordinator of the "Food quality control" course delivered within the frame of graduate studies in Food Engineering hosted by the Faculty of Agronomy and Food Technology University of Mostar.

From 2012 on: Lecturer delivering tuition within the frame of undergraduate studies hosted by the Faculty of Food Technology and Biotechnology University of Zagreb; Course: Contemporary techniques employed in food quality control.

From 2011 on: Lecturer delivering tuition within the frame of postgraduate vocational studies in Food Quality & Safety hosted by the Faculty of Food Technology and Biotechnology University of Zagreb; Course: Analytical techniques employed in food safety monitoring.

From 2011 on: Lecturer delivering tuition within the frame of postgraduate vocational studies in Traditional meat products' technologies hosted by the Faculty of Food Technology "J J Strossmayer" University of Osijek; Course: Sensory and physicochemical features of traditional meat products.

Organisation of Scientific Meetings:

From 2017 on: Scientific Committees appointed by the Croatian Food Agency – Chairlady of the Scientific Committee in charge of food safety hazards of chemical nature, 7 members, Croatia.

From 2017 on: Science Councils of the Croatian Food Agency – member of the CFA Science Council in charge of food-related issues, 7 members, Croatia.

2015: - XI Poultry Days, member of the Scientific Committee, science/expert symposium with international participation, 100 members, Croatia.

2014: 8th International Congress of Food Technologists, Biotechnologists and Nutritionists, member of the Scientific Committee, 250 participants, Croatia.

2013 – 2016: Scientific Committees appointed by the Croatian Food Agency – member of the Scientific Committee in charge of food safety hazards of chemical nature, 7 members, Croatia.

Organisational Responsibilities:

From 2014 on: member of the task-force appointed by the European Commission (Brussels, Belgium) – representative of the NRL delegated to present national achievements in poultry meat quality improvement.

From 2010 on: auditor of the quality assurance systems established by laboratories entrusted with food and feed analytics.

From 2007 on: Head of the Laboratory for Analytical Chemistry of the Croatian Veterinary Institute Zagreb, designated as the national food quality & safety reference laboratory (the NRL for anabolic substances present in food of animal origin and the NRL for poultry meat).

From 2007 on: member of the Science Council of the Croatian Veterinary Institute Zagreb, Croatia.

Memberships:

From 2019 on: member of the Editorial Board of the science/expert journal Croatian Journal of Food Science and Technology.

From 2017 on: Chairlady of the Scientific Committee in charge of food safety hazards of chemical nature and a member of the Science Council of the Croatian Food Agency.

From 2017 on: member of the National Committee for Codex Alimentarius.

From 2016 on: representative of the Croatian Veterinary Institute in the Committee entrusted with drafting and preparation of standards applicable to Croatian public nutrition, the Healthy Meal Association

From 2016 on: member of the Editorial Board of the science/expert journal "Meso" ("Meat").

2014: member of the task-forces entrusted with the preparation of science study "Improvement of pig meat quality and safety: Frozen pig meat quality", Croatian Food Agency.

2013 – 2016: member of the Science Committee in charge of food safety hazards of chemical nature, Croatian Food Agency.

2012: member of the task-force entrusted with the preparation of Scientific Opinion addressing polyphosphates in meat and meat products, Croatian Food Agency.

From 2012 on: member of the Alumni & Friends Association of the Faculty of Food Technology Osijek.

From 2011 on: member of the Editorial Board of the science/expert journal "Veterinarska Stanica".

2011: Member of the task-force entrusted with the preparation of Scientific Opinion addressing frozen poultry meat quality, Croatian Food Agency.

From 2010 on: member of the Croatian Society of Food Technologists, Biotechnologists and Nutritionists established under the wing of the Faculty of Food Technology and Biotechnology University of Zagreb.

From 2009 on: member of a number of expert panels entrusted with the preparation of national ordinances and state monitoring implementation, as well as expert teams entrusted with the preparation and revision of, and expert opinions on, drafted regulations and other legal documents governing food quality and safety at the European Union level; the Ministry of Agriculture of the Republic of Croatia.

In: Ochratoxin A and Aflatoxin B1
Editor: Reuben Hess
ISBN: 978-1-53617-416-8

Chapter 3

CO-OCCURRENCE OF OCHRATOXIN A AND CITRININ AND IDENTIFICATION OF FUNGI THAT PRODUCE THE RESPECTIVE MYCOTOXINS IN MOLD-CONTAMINATED RICE, CORN AND GROUNDNUT SAMPLES

L. Gayathri[1,2,5], PhD, B. Karthiyayini[3], K. Ruckmani[1], PhD, D. Dhanasekaran[2], PhD and M. A. Akbarsha[4,5,*], PhD

[1]Department of Pharmaceutical Technology,
University College of Engineering, Anna University-BIT campus,
Tiruchirappalli,Tamil Nadu, India
[2]Department of Microbiology, Bharathidasan University,
Tiruchirappalli, Tamil Nadu, India
[3]National Centre for Alternatives to Animal Experiments,
Bharathidasan University, Tiruchirappalli, Tamil Nadu, India
[4]National College (Autonomous), Tiruchirappalli, Tamil Nadu, India

[*]Corresponding Author's Email: akbarbdu@gmail.com.

[5]Mahatma Gandhi-Doerenkamp Centre for Alternatives, Bharathidasan University, Tiruchirappalli,Tamil Nadu, India

ABSTRACT

OchratoxinA (OTA) and citrinin (CTN) are the most commonly co-occurring mycotoxins in a wide variety of food and feed commodities. As a combination these mycotoxins are reported to cause endemic nephropathy, hepatotoxicity and pulmonary toxicity. Hence, it is important to check the quality of food commodities that are destined for human and/or animal consumption, with special reference to occurrence/co-occurrence of mycotoxins. Therefore, in this study a total of sixty mold-contaminated rice, corn and groundnut samples were collected from one of the Cauvery delta regions of Tamil Nadu, India, and analyzed to find the co-occurrence of OTA and CTN, and the respective mycotoxigenic fungal strains that produce these toxins. Co-occurrence of OTA and CTN was qualitatively and quantitatively confirmed by thin layer chromatography (TLC) and High Pressure Liquid Chromatography (HPLC) methods. OTA and CTN co-occurred in 35%, 25% and 15% of the contaminated rice, corn and groundnut samples, respectively. From the mold-contaminated rice, corn and groundnut samples, a total of 60 morphologically distinct strains of *Aspergillus* sp and *Penicillium* sp were isolated and pre-screened for the production of mycotoxin(s) using coconut cream agar (CCA) plating method. Isolates of fungi that fluoresced highly were cultured in yeast extract sucrose (YES) broth and the production of OTA and CTN was analyzed using HPLC. OTA- and CTN-positive strains were characterized and identified by ITS gene sequence analysis as *Aspergillus tubingensis, A. flavus, A. niger,* and *A. oryzae*. Co-occurrence of OTA & CTN and identification of mycotoxigenic fungi that produce the respective mycotoxins present in rice, corn and groundnut indicate the potential health risk of these mycotoxins either singly or in combination to humans as well as animals.

Keywords: Mycotoxins, ochratoxin A, citrinin, mycotoxin-producing fungi, HPLC

ABBREVIATION

OTA	Ochratoxin A
CTN	Citrinin
CCA	coconut cream agar
YES	yeast extract sucrose
HPLC	High Pressure Liquid Chromatography
CTAB	Cetyltrimethyl ammonium bromide
UV	Ultra-violet
EDTA	Ethylene diamine tetra acetic acid;
PCR	Polymerase Chain Reaction.

1. INTRODUCTION

Mycotoxins are secondary metabolites produced by fungi that contaminate food and feed commodities either in the field or during storage. At some stage in storage, fungal attack is probably influenced by many biotic and abiotic factors (Molinie et al., 2005; Pitt & Hocking 2009). Upon contamination, a fungal species becomes capable of producing more than one mycotoxin or several mycotoxins are produced by different fungal species which co-infest the food/feed commodity. Therefore, often more than one mycotoxin is found on contaminated substrate of foodstuffs (Molinie et al., 2005; Reddy et al., 2010; Streit et al., 2012). Thus, as a mixture these mycotoxins cause greater adverse health effects to humans than a single mycotoxin.

OTA and CTN are quite common mycotoxins as they are produced by *Aspergillus* and *Penicillium* which are worldwide in distribution (Nguyen et al., 2007; Reddy et al., 2010; Klaric et al., 2013). *A. ochraceus, A. carbonarius, A. niger, P. verrucosum, P. expansum, P. griseofulvum* and *P. citrinum,* predominantly found in cereal grains, cocoa, spices, oilseeds, coffee beans and legumes, were reported to produce these mycotoxins (Pitt, 1987; Abraca et al., 1994;; Wolff, 2000; Serra et al., 2003). As a result, these mycotoxins can co-occur in a wide variety of food commodities. In fact several studies have shown co-occurrence of OTA and CTN in agricultural crops, foods and feed commodities (Jelinek et al., 1988; Nguyen, 2007). Experimental evidences for the carcinogenic potential of OTA led to its classification by the International Agency for Research on Cancer (IARC) as a potential human carcinogen under group 2B whereas CTN is classified under group 3 in view of inadequate evidences of *in vivo* carcinogenicity (IARC, 1993). Many adverse effects of mycotoxins, OTA and CTN individually as well as in combination, were reported which include endemic nephropathy, hepatotoxicity and pulmonary toxicity (Follmann et al., 2000; Bouslimi et al., 2008; Gayathri et al., 2015). Even OTA and CTN have been proved to be stable without any degradation when cooked at high temperature (Osborne et al., 1996; Kabak, 2009). Thus, it is pertinent to study the co-occurrence of OTA and CTN in long-time stored foods and feed commodities that are destined for human and/or animal consumption.

India is an agrarian country and most of the people depend on agriculture for livelihood. The productivity of food grains in India is high, but then unscientific/improper storage of food grains and commodities leads to fungal infestation and mycotoxin production. The Food and Agricultural Organization (FAO) has reported that 25% of agricultural crops worldwide are contaminated by mycotoxins (WHO, 1999; Wagacha and Muthomi, 2008). Rice, corn and groundnut are the most important staple foods used around the world and they have also been used as livestock feed and as raw material for the production of breakfast cereals, corn chips, grits, flour, oil and industrial starch.

Studies regarding distribution of mycotoxigenic fungi and the contamination of mycotoxins, apart from aflatoxin B1, in stored food grains are rare, especially in India. Regulatory limit has been prescribed for aflatoxin B1 in India but the other mycotoxins can also contaminate the food commodities and cause diseases to Indian population (Bhat et al., 1996, 1997; Janardhana et al., 1999; Kishore et al., 2002; Somashekar et al., 2004; Gautam et al., 2012). Therefore, the present study was undertaken to obtain preliminary information about co-occurrence of ochratoxin A and citrinin, and the toxigenic fungi that produce these toxins, in rice, corn and groundnut collected from Tiruchirappalli district of India, which is one of the Cauvery Delta regions of Tamil Nadu.

2. Materials and Methods

2.1. Sample Collection

A total of sixty randomly contaminated samples of rice (n = 20), corn (n = 20) and groundnut (n = 20) were collected during winter season (December 2014 to January 2015) from the local markets of Tiruchirappalli district of Tamil Nadu, India. The minimum sample size of 250 g was collected and kept at -20° C while awaiting analysis. Each sample was separated to equal size to find the co-occurrence of OTA and CTN and to detect the OTA- and CTN-producing fungi in contaminated rice, corn and groundnut samples.

2.2. Reagents

OTA and CTN standards were obtained from Sigma Chemical Company (St. Louis, MO, USA). Standard solutions were prepared by dissolving 5 mg of OTA and CTN, separately, in 1 mL of chloroform. All reagents and solvents (hydrochloric acid, chloroform, acetonitrile, *n*-hexane, methanol, propanol-2-ol, acetic acid, toluene, oxalic acid and ethyl acetate, and potassium chloride) were of HPLC grade, and purchased from HiMedia, Mumbai, India.

2.3. Screening of OTA, CTN and OTA+CTN by Thin Layer Chromatography (TLC)

Crude extract of the sample was obtained by mixing 100 g of the powdered sample and 100 mL of chloroform in an orbital shaker, followed by filtration through Whatman No 1 filter paper. An equal volume of chloroform was added to the filtrate and again mixed for 30 min in an orbital shaker. The organic phase in chloroform was collected and dried under vacuum by using a rotary evaporator in a 40° C water bath at low speed. The powder was dissolved in chloroform and spotted as 10 μL drops on to a 0.5 mm thick silica gel coated on a glass plate and dried. The standard solutions of OTA and CTN were also spotted on to the same plate at 3 μg concentration. The plate was developed in mobile phase condition of toluene: ethyl acetate: formic acid (6:3:1) by the revised method of Ramesh et al. (2013). After development, the plate was air-dried and observed under UV light at 366 nm. The fluorescence intensities of OTA (blue) and CTN (lemon yellow) spots of mold-contaminated rice, corn and groundnut samples were compared with the standard spots. Standard was also used to compare the color and Rf value of the test sample on the plate. The concentrations of OTA and CTN were determined based on the concentrations of standard toxins. Samples that were positive for OTA and CTN co-occurrence were further confirmed and quantified adopting HPLC.

2.4. Extraction, Confirmation and Quantification of Co-Occurrence of OTA- and CTN- in Mold-Contaminated Rice, Corn and Groundnut Samples

2.4.1. Extraction of OTA and CTN

Extraction, purification and HPLC conditions were as prescribed by Nguyen et al. (2007). One hundred grams of powders of contaminated rice, corn and groundnut were extracted with 100 mL of acetonitrile containing 4% aqueous solution of potassium chloride (9:1). The PH of the mixture was adjusted to 1.5 with undiluted hydrochloric acid, and then shaken on an orbital shaker for 20 min and filtered through a Whatman # 1 filter paper under vacuum.

2.5. Purification

One hundred mL of n-hexane was added to the filtrate and shaken for 10 min. After separation, the upper phase (n-hexane) was discarded. This step was repeated with 50 mL of n-hexane. To the lower phase, 50 mL de-ionized water and 50 mL chloroform were added. The solution was shaken for 10 min. After separation, the lower phase (chloroform) was collected. The upper phase was re-extracted, twice, with 25 mL of chloroform each time. The chloroform extracts were pooled and evaporated to near dryness under vacuum using a rotary evaporator in a 40℃ water bath at low speed. The dried sample was weighed and re-dissolved in chloroform for HPLC analysis.

2.6. HPLC Analysis

A Shimadzu (Kyoto, Japan) LC-8A system, equipped with an injector 20 μL loop, a C18 column (Phenomenex) and a UV visible detector model SPD-20 A, was used. Different excitation and emission fluorescence parameters (OTA 335 and 465 nm; CTN 331 and 500) were used to detect each extract. The system was run with acetic acid/acetonitrile/propan-2-ol (65/40/5) as the mobile phase; the flow rate was 0.5 mL/min; elution time of OTA and CTN were about 5.2 and 6.1 min, respectively. The chromatogram was analyzed by LC solution real time analysis software.

2.7. Isolation and Molecular Characterization of OTA- and CTN-Producing Fungal Isolates

2.7.1. Isolation of Fungi

Serial dilution technique was adopted for isolation of fungi from the mold-contaminated rice, corn and groundnut. In this technique, suspension of the sample was prepared by adding 1 g powder of the sample to 10 mL of distilled water and shaking for 15 min in an orbital shaker. Thereafter, each suspension was serially diluted. From 10^{-3}, 10^{-4}, 10^{-5} dilutions, 0.1 mL was placed on potato dextrose agar plates by spread plate method and the plates were incubated in sterile moist chamber at 28° ± 2°C for 7 days until fungal colonies appeared. Chloramphenicol (0.03 mg/L) was added to the medium as antibiotic to avoid bacterial contamination. The fungal colonies were identified based on colony morphology and microscopic features, and stored at -20 °C until use.

2.7.2. Pre-Screening of Mycotoxin-ProducingFungal Isolate

In order to find the mycotoxin-producing fungi from the total isolate, coconut cream agar (CCA) plating method was adopted (Dyer and McCammon, 1994; Heenan et al., 1998). From all the fungi that were isolated, only the morphologically distinct fungal isolates were chosen and inoculated on CCA and incubated at 28 °C for 7 days, to examine the mycotoxin-producing fungal isolates. When the plates were illuminated in an UV chamber at 366 nm, the fluorescent zones formed by the fungal colonies were used to select the putative mycotoxin-producing isolates.

2.7.3. Solvent Extraction of Mycotoxin-ProducingFungal Isolates

The highly fluorescing fungal isolates were selected and separately inoculated into 500 mL of yeast extract sucrose broth (YES) according to Abarca et al. (1994). Cultures of fungal isolates were grown on YES broth at 25 °C for 7 days, after which the filtrate was separated out with Whatman # 1 filter paper, to collect the filtrate in a flask. An equal volume of chloroform was added and shaken for 30 min. A separating funnel was used to collect the lower phase in a conical flask and, again, an equal volume of chloroform was added and shaken for 30 min. The lower phase was pooled and evaporated to near dryness under vacuum by using a rotary evaporator maintained at 40℃ in a water bath, at low speed. Thereafter, the precipitate was again dissolved in chloroform and the samples were subjected to HPLC analysis. Chromatograms of test samples were compared with standard toxins. OTA- and CTN-producing fungal isolates were chosen for the molecular characterization.

2.8. Extraction and Analysis of Genomic DNA

2.8.1. Extraction

The total genomic DNA of OTA- and CTN-producing fungal isolates AGR11, AGR12, AGC2, AGC8, AGG2 and AGG4 were extracted by the modified CTAB (Cetyl-Trimethyl Ammonium Bromide) method described by Gullino and Garibaldi (2010). At first, 50 mg of mycelia were scraped from 10 day old PDA culture and manually pulverized by adding 500 μL of TES lysis buffer (100 mMTris pH 8.0, 10 mM EDTA and 2% SDS) pre-warmed at 60 ºC. Then, 50 μg of proteinase K was added to the suspension and incubated at 60 ºC for 60 min with intermittent mixing. The sample was cooled to room temperature, and 140 μL of 5 M NaCl and 64 μL 10% CTAB (w/v) were added. Further, the sample was incubated at 65 ºC for 10 min, and equal volumes of chloroform and isoamyl alcohol (24:1) were added. The preparation was centrifuged at 14000 rpm for 10 min and the upper aqueous layer was carefully transferred to a new tube. DNA was then precipitated using 600 μL cold isopropanol and 100 μL 3 M sodium acetate (pH 5.2) and maintained at -20 ºC. The DNA thus precipitated was washed twice with 70% ethanol and allowed to dry in air under sterile conditions. Finally, the pellet was resuspended in 100 μL of TE buffer (10 mMTris-HCl, pH 7.5, and 1 mM EDTA, pH 8.0) and stored at -20 °C for further analysis. The quality of the extracted DNA was verified by gel electrophoresis on 0.8% agarose containing ethidium bromide, and observed in a gel documentation system (Bio-Rad, USA).

2.8.2. PCR Amplification

The internal transcribed sequence (ITS) of the DNA was amplified in an automated thermal cycler (Eppendorf) using the universal primers, viz., ITS1 (5′-TCCGTAGGTGAACCTGCGG-3′) and ITS4 (5′TCCTCCGCTT AT TGATATGC-3′) as described by White et al. (1990). The reaction was executed in a 30 μL volume containing 15 μL 1 x PCR premix, 2 μL template DNA, 0.5 μL each primer (10 μM) and 12 μL distilled water. Cycling parameters were as follows: initial denaturation at 95°C for 2 min; 30 cycles of denaturation at 95°C for 2 min; annealing at 55°C for 30 sec;

extension at 72°C for 90 sec; and then final elongation at 72°C for 10 min. The amplified product was run on 1.2% agarose gel in TAE buffer (40 mMTris base, pH 7.6, 20 mM acetic acid, and 1 mM EDTA, pH 8.0) to ensure the quality.

2.8.3. Sequencing and Phylogenetic Analysis

The amplicon was purified at the DNA Synthesis and Sequencing Facility, Macrogen (Eurofins Genomics India Pvt. Ltd., Bangalore, India) using a PCR product pre-sequencing kit and sequenced on ABI 3730 x 1 automated sequencer (Perkin Elmer Applied Biosystems, Foster City, CA, USA) using 'ABI Prism BigDye Terminator Cycle Sequencing Ready Reaction Kit' with AmpliTaq® DNA polymerase, according to the protocol recommended by the manufacturer. Single-pass sequencing was performed on each template using the amplification primers. The acquired electropherograms were analysed and sequence contigs were assembled and edited using Codon code Aligner v4.1.1. Further, BLASTN search was conducted for sequence similarity. Multiple sequence alignment was carried out using ClustalW. Concurrently, the phylogenetic tree was constructed by neighbor-joining method adopting MEGA 6.0 software (Tamura et al., 2011), where the interior branch lengths were tested via bootstrap analysis with 1000 replications and the sequences were deposited in NCBI database by BankIt software tool available online (http://www.ncbi.nlm.nih.gov).

3. RESULTS

3.1. Screening, Confirmation and Quantification of Co-Occurrence of OTA and CTN in Rice, Corn and Groundnut

The crude extracts obtained from contaminated rice, corn and groundnut were subjected to TLC in order to screen and quantify OTA or CTN and their co-occurrence (Figures 1-3). Out of the 20 rice samples, 30%, 15% and 35%, respectively, were contaminated with OTA or CTN alone and also OTA+CTN. In corn, OTA alone was found to contaminate 60% of the samples, and CTN was found to co-occur with OTA in 25% of the samples.

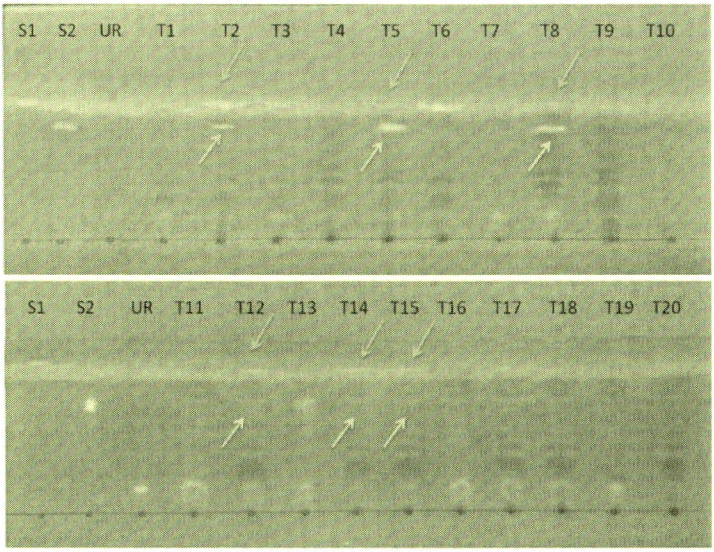

Figure 1. Thin layer chromatograms showing occurrence of OTA or CTN alone and as OTA+CTN in the crude extract obtained from rice samples. Lane S1: OTA standard; S2: CTN standard: UR: Crude extract of uncontaminated rice sample; T1 to T20: Crude extracts of contaminated rice samples. Arrow-marks indicate co-occurrence of OTA and CTN.

The frequency of OTA and CTN contamination in groundnut was lesser than in rice and corn samples. Only 15% of the groundnut samples were contaminated with OTA and/or OTA+CTN whereas CTN alone was found to contaminate only 5% of the samples. In order to confirm the co-occurrence of OTA and CTN using a valid HPLC method, rice (T2, T5, T8, T12, T14, T15, T17), corn (T2, T3, T8, T9, T10) and groundnut (T2, T3, T9) samples positive for the co-occurrence of OTA and CTN, identified from TLC, were again taken for the simultaneous extraction of OTA and CTN followed by purification. The analytical HPLC chromatogram of standard OTA was matched with the chromatogram of OTA-positive extracts (Figures 4-6), and the chromatogram of standard CTN was matched with the chromatogram of CTN-positive extracts of rice, corn and groundnut samples (Figures 7-9). The occurrence of OTA or CTN and co-occurrence of OTA and CTN were evaluated and quantified in relation to standard mycotoxins and the results are consolidated in Table 1.

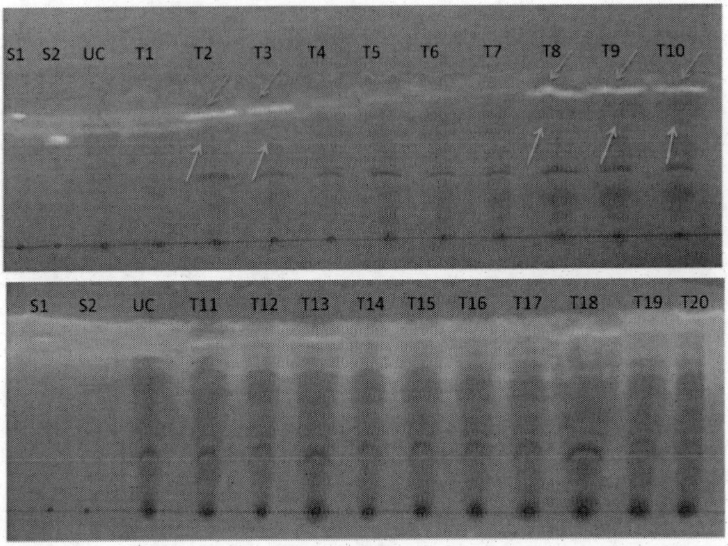

Figure 2. Thin layer chromatograms showing occurrence of OTA or CTN alone and as OTA+CTN in the crude extract obtained from corn samples. Lane S1: OTA standard; S2: CTN standard: UC: Crude extract of uncontaminated corn sample; T1 to T20: Crude extracts of contaminated corn samples. Arrow-marks indicate co-occurrence of OTA and CTN.

Table 1. The average and range of OTA and CTN alone as well as in combination in rice, corn and groundnut samples

Sample	No. of samples Analyzed	Mycotoxin	Detectable samples	Quantification of toxins	
				Average µg/kg	Range of toxin µg/kg
Rice	20	OTA	5/20	5.75	1.5-10
		CTN	3/20	1.25	0.5-2
		OTA+CTN	7/20	7.5+7.1	<5-10 + 6-8.25
Corn	20	OTA	12/20	9.3	6.7-12
		CTN	-	-	-
		OTA+CTN	5/20	8.5+ 4.25	<6 -11 + 3.5-5
Groundnut	20	OTA	3/20	5.4	4.3-6.5
		CTN	1/20	4.2	4.2
		OTA+CTN	3/20	5.95+5.1	<4.9 -7 + 4.2-6

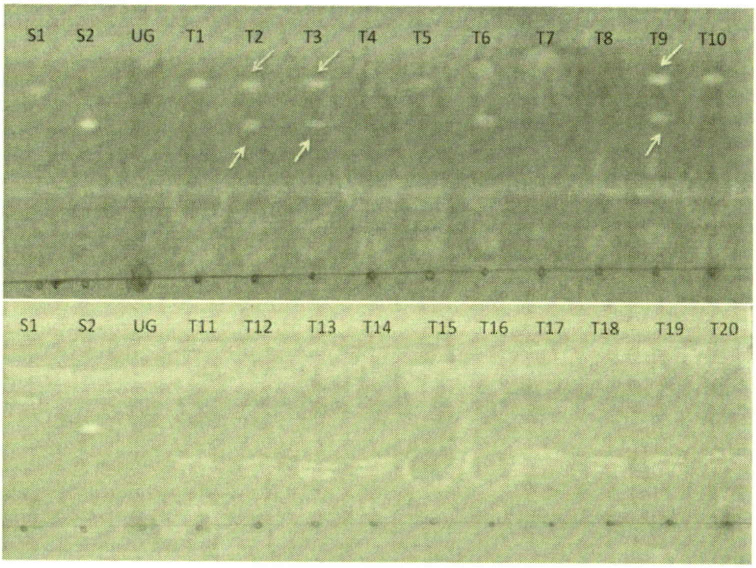

Figure 3. Thin layer chromatograms showing occurrence of OTA or CTN alone and as OTA+CTN in the crude extract obtained from groundnut samples. Lane S1: OTA standard; S2: CTN standard: UG: Crude extract of uncontaminated groundnut sample; T1 to T20: Crude extracts of contaminated groundnut samples. Arrow marks indicate the co-occurrence of OTA and CTN.

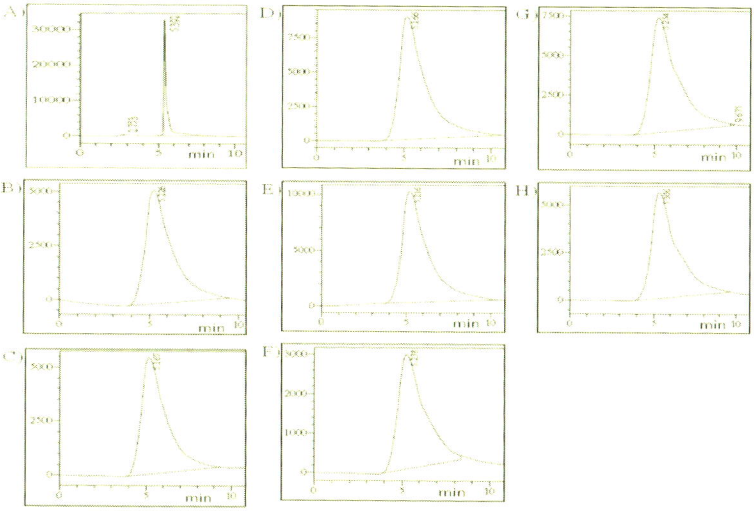

Figure 4. Chromatograms of OTA separation at 335 nm and 465 nm wavelength by HPLC. (A) OTA standard. (B-H) Crude extracts of contaminated rice samples showing the presence of OTA.

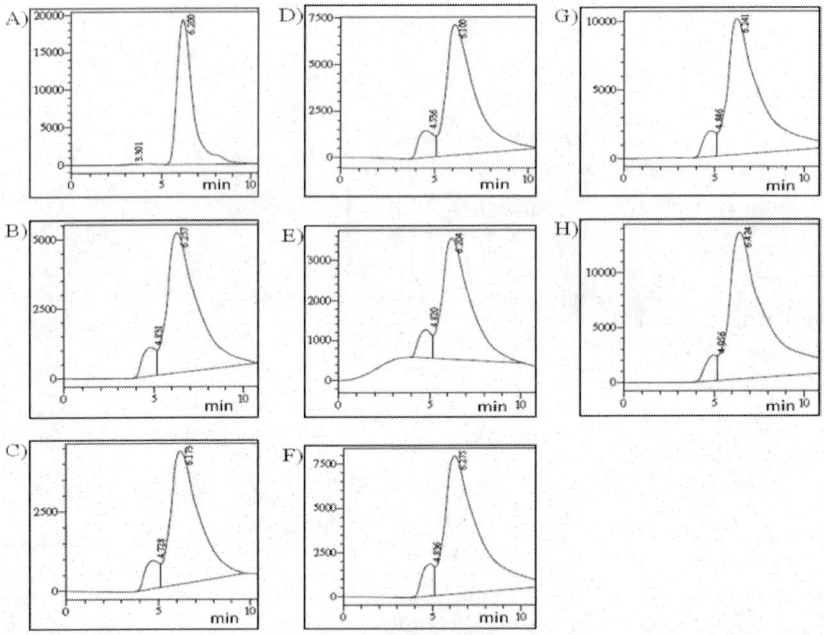

Figure 5. Chromatograms of CTN separation at 331 nm and 500 nm wavelength by HPLC. (A) CTN standard. (B-H) Crude extracts of contaminated rice samples showing the presence of CTN.

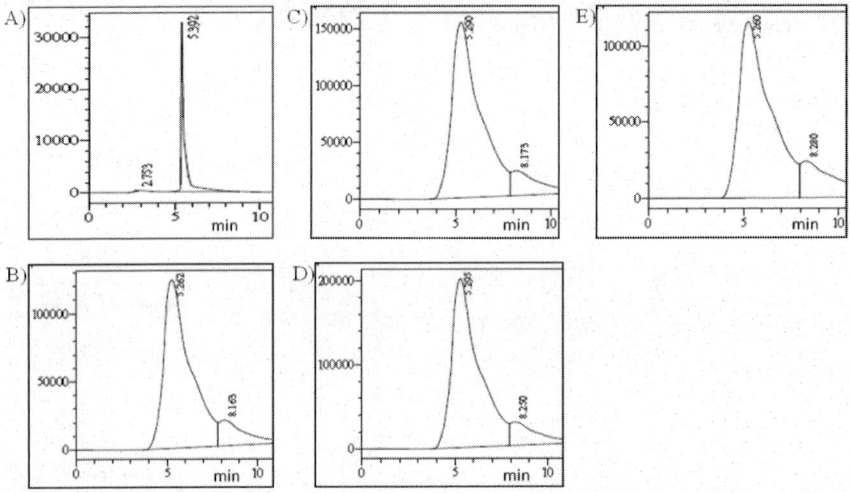

Figure 6. Chromatograms of OTA separation at 335 nm and 465 nm wavelength by HPLC. (A) OTA standard. (B-D) Crude extracts of contaminated corn samples showing the presence of OTA.

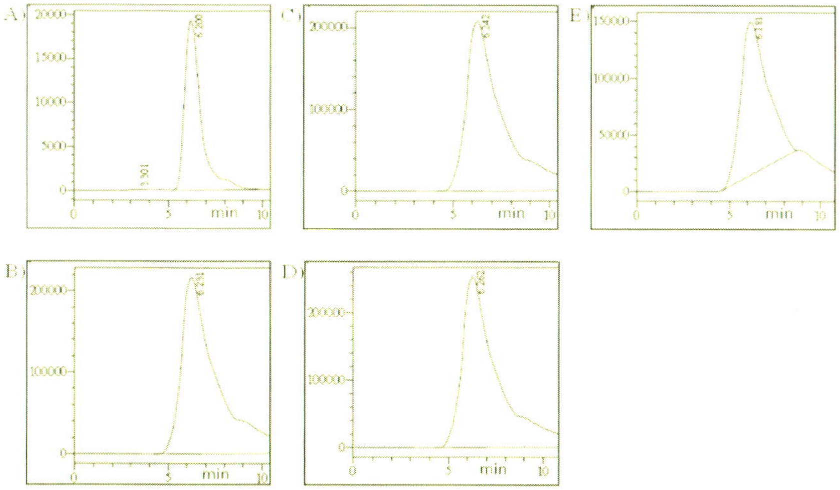

Figure 7. Chromatograms of CTN separation at 331 nm and 500 nm wavelength by HPLC. (A) CTN standard. (B-D) Crude extracts of contaminated corn samples showing the presence of CTN.

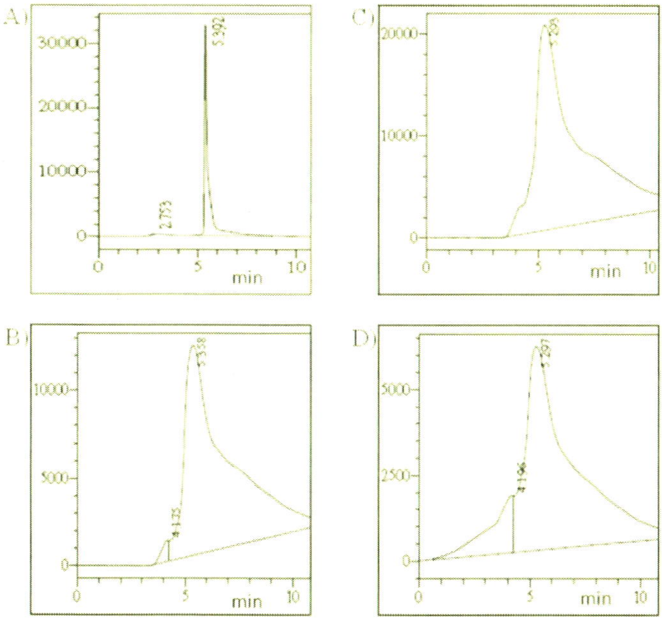

Figure 8. Chromatograms of OTA separation at 335 nm and 465 nm wavelength by HPLC. (A) Chromatogram of OTA standard. (B-D) Chromatograms of crude extracts of contaminated groundnut samples showing the presence of OTA.

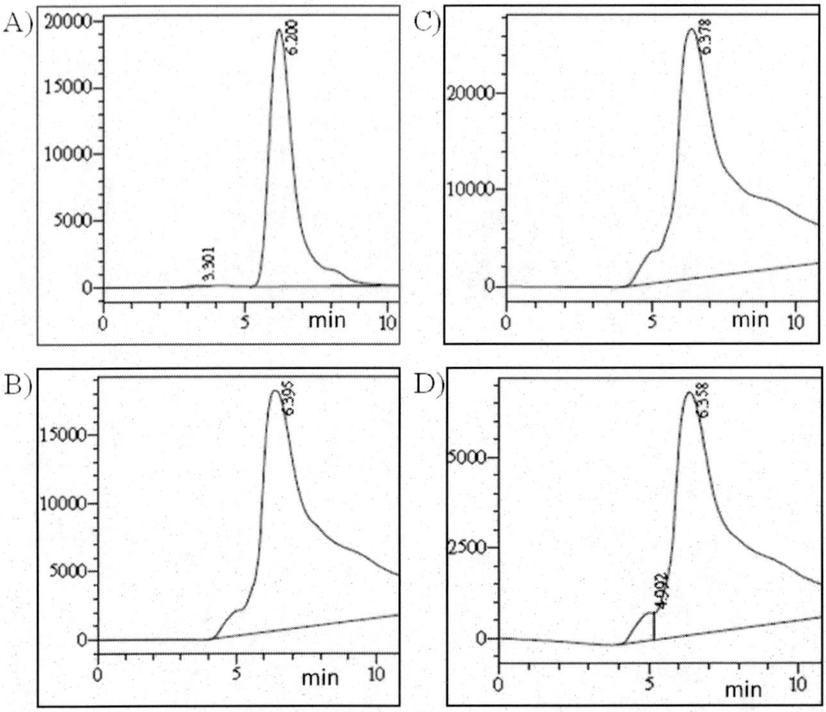

Figure 9. Chromatograms of CTN separation at 331 nm and 500 nm wavelength by HPLC. (A) CTN standard. (B-D) Crude extracts of contaminated corn samples showing the presence of CTN.

3.2. Isolation of Mycoflora from Contaminated Rice, Corn and Groundnut

The contaminated rice, corn and groundnut samples were used for the isolation of mycotoxigenic fungi in order to find the OTA- and CTN-producing fungal isolates. All the sixty samples had different levels of fungal contamination and viable count of mycoflora was in the range 2.5-11.0 cfu x 10^5/g. A total of 147 fungal isolates were obtained from rice, corn and groundnut samples, from which the morphologically distinct 60 fungal isolates were selected for identification of the genus. Totally, 35 isolates of *Aspergillus* sp and 25 isolates of *Penicillium* sp were identified from rice, corn and groundnut samples. Incidence of *Aspergillus* sp. was found to be high in all the food samples (Figure 10).

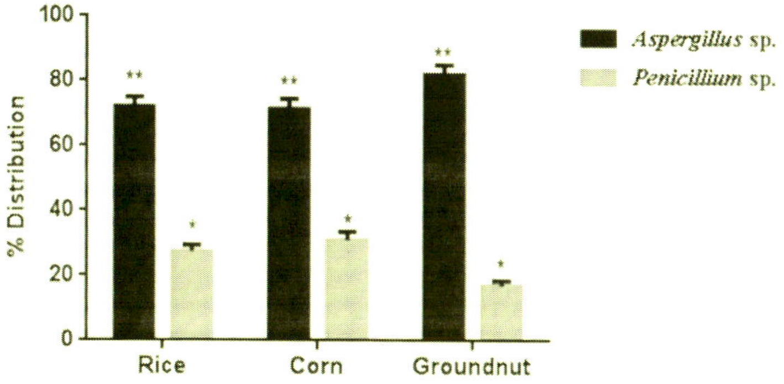

Figure 10. Relative distribution of *Aspergillus* and *Penicillium* sp in rice, corn and groundnut samples. Data are expressed as mean ± SD of three independent experiments. *P< 0.05; **P< 0.01 compared to control.

3.3. Preliminary Screening of Mycoflora for Mycotoxin Production

Out of the 60 morphologically distinct *Aspergillus* and *Penicillium* isolates, 30%, 37% and 64% of mycotoxin-producing *Aspergillus* and 6%, 6% and 4% of mycotoxin-producing *Penicillium* species were identified in rice, corn and groundnut, respectively. Therefore, the preliminary screening for isolated *Aspergillus* and *Penicillium* species on CCA revealed that mycotoxin-producing *Aspergillus* sp were found to be highest in groundnut samples compared to rice and corn whereas mycotoxin-producing *Penicillium* sp were minimal in rice, corn and groundnut (Figure 11). Based on their fluorescent zone on the 7th day of pure culture, the fungal isolates were categorized into three groups, viz., low, medium and high (Figure 12). Accordingly, the toxigenic fungal isolates of rice, corn and groundnut were graded and listed in Table 2. Out of all toxigenic fungal isolates, only the highly fluorescing fungal isolates were selected to find if they produce either OTA or CTN or both.

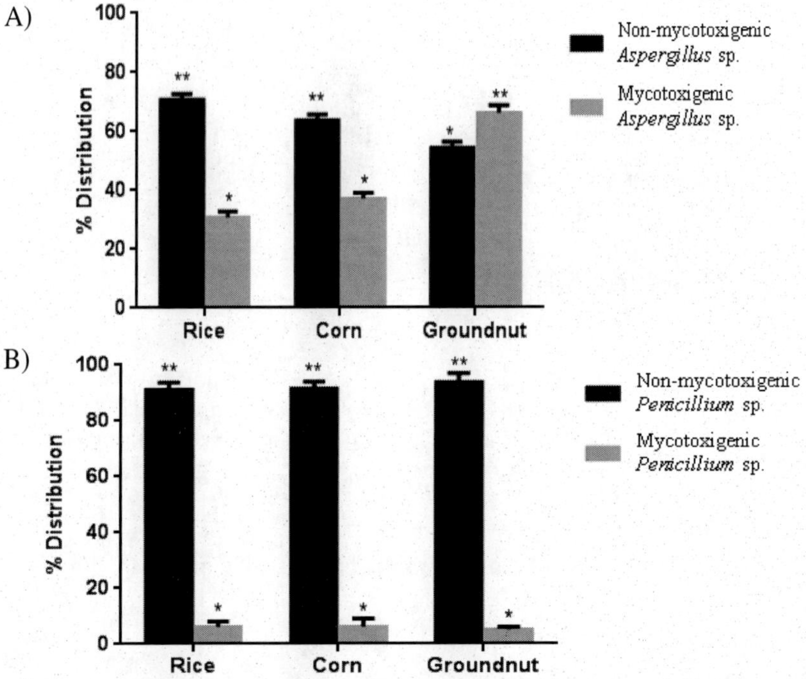

Figure 11. Distribution of toxin producing *Aspergillus* and *Penicillium* sp. in rice, corn and groundnut. (A) Distribution of non-mycotoxic and mycotoxicogenic *Aspergillus* sp. (B) Distribution of non-mycotoxic and mycotoxicogenic *Penicillium* sp.

Table 2. Categorization of mycotoxigenic fungal isolates based on the difference in the fluorescence on coconut cream agar plate obtained from rice, corn and groundnut samples

Mycotoxigenic fungal isolates obtained from rice			
Fungal Isolates	Low	Medium	High
PGR1		++	
AGR5		++	
AGR6			+++
AGR10	+		
AGR11			+++
AGR12	+		+++
Mycotoxigenic fugal isolates obtained from corn			
AGC1			+++
AGC2			+++

Mycotoxigenic fugal isolates obtained from corn			
AGC5		++	
AGC6			+++
AGC7			+++
AGC8			+++
AGC9			+++
AGC11		++	
AGC12		++	
AGC16	+		
PGC20		++	
PGC21		++	
Mycotoxigenic fugal isolates obtained from groundnut			
AGG1		++	
AGG2			+++
AGG3			+++
AGG4			+++
AGG5			+++
AGG6			+++
AGG7			+++
AGG13		++	
AGG15	+		
AGG19	+		
AGG20	+		
PGG21	+		
PGG22		++	

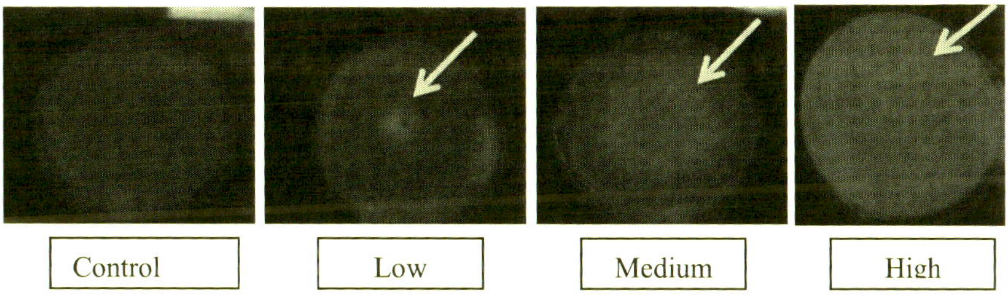

Figure 12. Representative photographs for the difference in the fluorescent emission of toxigenic fungal isolates on CCA agar after 7 days of incubation.

3.4. Screening of Fungal Isolates for Mycotoxin Production Adopting HPLC

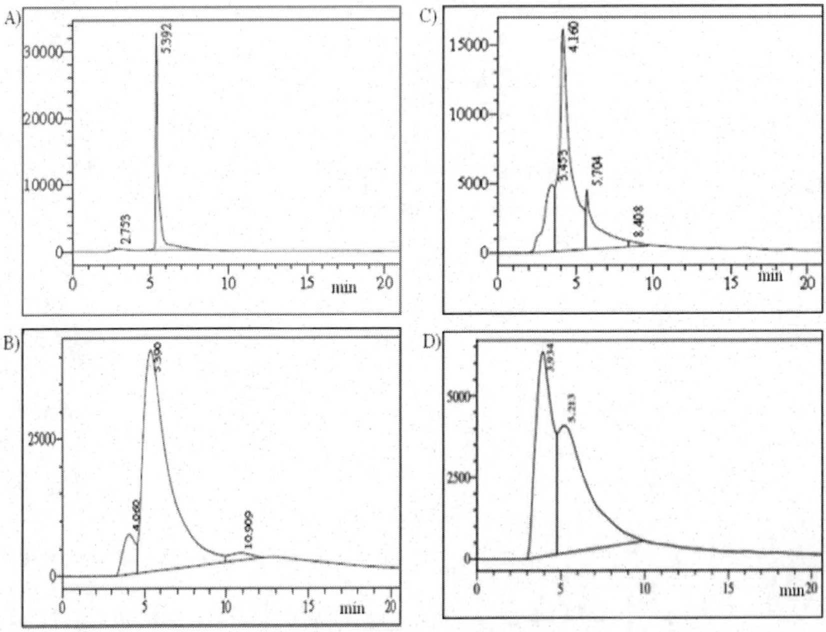

Figure 13. Chromatograms of OTA and crude extract obtained from fungal isolates. (A) OTA standard. (B-D) Crude extracts obtained from AGR12 (B), AGC2 (C), AGG2 (D) were positive for OTA.

Based on the CCA screening, the most highly fluorescing 3 fungal isolates (AGR6, AGR11 and AGR12) from rice, 6 fungal isolates (AGC1, AGC2, AGC6, AGC7, AGC8 and AGC9) from corn and 6 fungal isolates (AGG2, AGG3, AGG4, AGG5, AGG6 and AGG7) from groundnut samples were grown in YES medium. The chloroform extracts were recovered and the samples were subjected to HPLC analysis to confirm the ability of the fungi to produce either ochratoxin A or citrinin or both. Chromatograms of standard toxins and the OTA- and CTN-positive fungal isolates are presented in Figures 13 and 14. The results revealed that, out of all the fungal isolates AGR11, AGC2 and AGG2 were positive for production of both ochratoxin A, and AGR12, AGC8, AGG4 were positive for the production of citrinin.

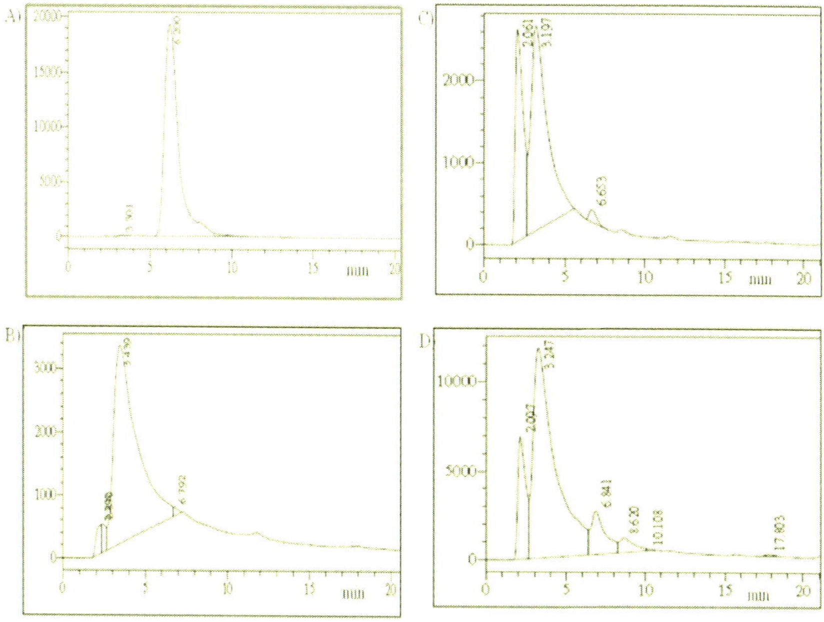

Figure 14. Chromatograms of CTN and crude extract obtained from fungal isolates. (A) CTN standard. (B-D) Crude extracts obtained from AGR11 (B), AGC8 (C), AGG4 (D) were positive for CTN.

3.5. Morphological and Molecular Characterization of OTA- and CTN- Producing Fungal Isolates

Figure 15 and Table 3 show the morphological characterization of fungal isolates AGR11, AGR12, AGC2, AGC8, AGG2, and AGG4. Fungal isolates that indicated confirmation to produce OTA and CTN were used to isolate the DNA for PCR amplification of internal transcribed region with the universal primer ITS1 and ITS4. An amplification product of ~600 bp was obtained for all isolates as shown in Figure 16. The amplified region was sequenced and selected for characterization and identification of fungi. On the basis of gene similarity in NCBI database the isolates AGR11 and AGR12 from rice has 96% sequence homology with the *A. tubingenesis* (KT291440) and *A. flavus* (KT291437), respectively.

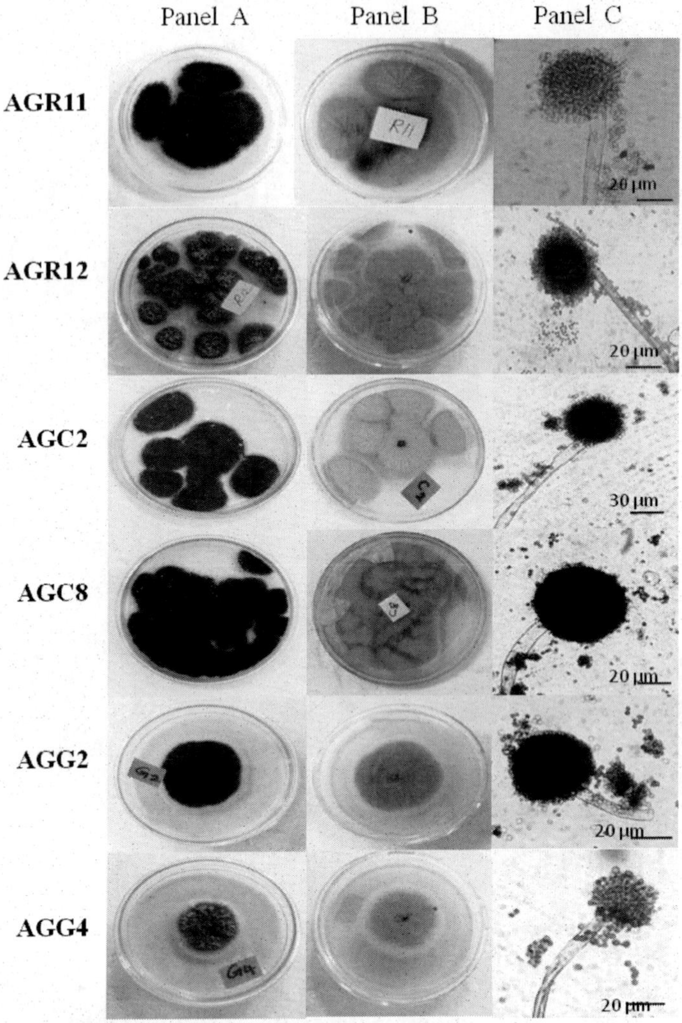

Figure 15. Cultural and morphological characteristics of OTA- and CTN- producing fungal isolates AGR11 and AGR12 obtained from rice, AGC2 and AGC8 obtained from corn, and AGG2 and AGG4 obtained from groundnut. Panel A- Macroscopic front view; panel B- Macroscopic back view of isolates and panel C- Lactophenol cotton blue stained conidial structures for fungi.

Isolates from corn, AGC2 and AGC8, showed 95% sequence homology with *A. tubingenesis* (KT291436) and *A. niger* (KT291438), respectively. Fungal isolates from groundnut, AGG2 and AGG4, showed 96% sequence homology with *A. niger* (KT291435) and *A. oryzae* (KT291439), respectively. Figure 17 shows the phylogenetic tree of fungal isolates constructed adopting neighbor joining method.

Table 3. Features of fungal isolates obtained from rice, corn and groundnut based on the morphological characterization

Fungal isolates	Colony morphology	Microscopic observation		Probable organism
		Conidia	Spores	
AGR11	Dark-brown to black	Conidial heads large, globose, dark brown	Smooth-walled	*Aspergillus* sp.
AGR12	Yellow-green	Conidial heads typically radiate, later splitting to form loose columns	Hyaline and coarsely roughened	*Aspergillus* sp.
AGC2	Black	Conidial heads large, globose, dark brown	Smooth-walled	*Aspergillus* sp.
AGC8	Dark-brown to black	Conidial heads large, globose, dark brown	Smooth-walled	*Aspergillus* sp.
AGG4	Black	Conidial heads large, globose, dark brown	Smooth-walled	*Aspergillus* sp.
AGG8	Yellow-green	Conidial heads radiate, conidia globose to subglobose	Long, coarsely roughened	*Aspergillus* sp.

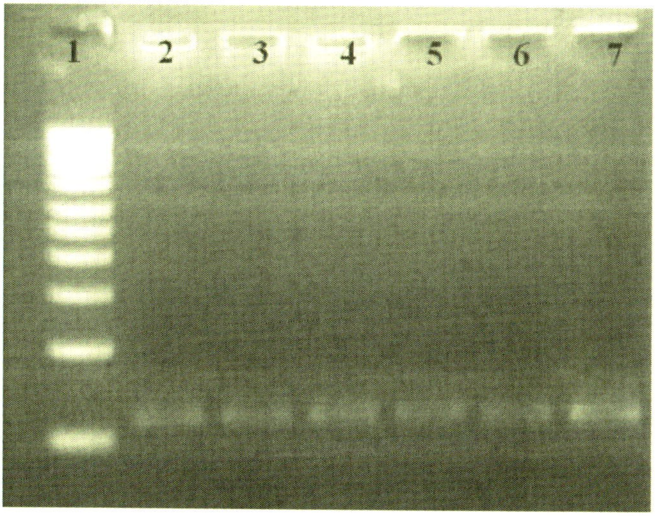

Figure 16. PCR amplification of ITS1 and ITS4 genes. Lane 1: 1Kb ladder; Lane 2: AGR11; Lane 3: AGR12; Lane 4: AGC2; Lane 5: AGC8; Lane 6; AGG2; Lane 7; AGG4.

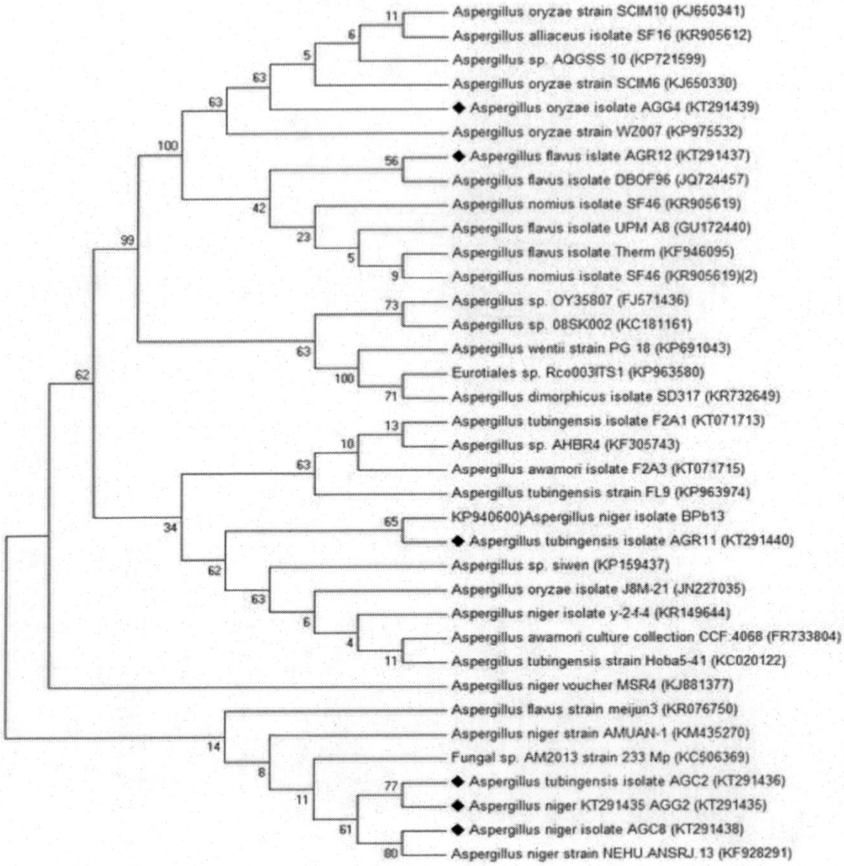

Figure 17. Phylogenetic tree analysis of OTA- and CTN-producing fungi by neighbor joining method using MEGA software. (Fungal strains identified in our study are indicated by diamond shape)

4. Discussion

Based on the climatic conditions, India is close to one of the high risk regions of mycotoxin contamination worldwide. In India Tamil Nadu, Kerala, Karnataka and Gangetic plain are highly vulnerable for the mycotoxin contamination (Bhat et al., 1978). Among different localities in India, Tamil Nadu is one of the leading producers of rice, corn and groundnut (Barah, 2009; Kumar et al., 2009). The tropical climate of Tamil Nadu is highly conducive for fungal contamination during pre- and post-harvesting seasons. Hence, stored food commodities can be contaminated by multiple fungal species and, therefore, the mycotoxins can be expected to co-occur. Considering concurrent contamination of mycotoxins and their toxicological risks, in this study we have evaluated the co-occurrence of OTA and CTN and OTA- and CTN-producing fungi in rice, corn and groundnut samples collected from Tiruchirappalli district (Cauvery delta region) of Tamil Nadu, India.

Among the various samples tested, co-contamination of OTA and CTN was found to be high in rice compared to corn and groundnut. In corn the contamination of OTA alone was found to be higher than OTA+CTN and in groundnut the prevalence of OTA and CTN alone as well as in combination was less. The possible reason behind the differential contamination of OTA and CTN alone as well as in combination in rice, corn and groundnut is connected with infestation of toxigenic fungi. The change in the substrate for the fungal growth and climatic conditions might influence the synthesis of mycotoxins on the chosen food types (Bhat and Vasanthi, 2003). Based on our results, we could conclude that rice, corn and groundnut samples collected from Tiruchirappalli district of Tamil Nadu are likely to be contaminated by OTA and CTN alone as well as in combination.

Data from quantification of OTA and CTN contamination in rice, corn and groundnut samples were matched with previous studies (Jackson and Ciegler, 1978; El-Maghraby and El-Maraghy, 1987; Puntarić, et al., 2001; Thirumala Devi et al., 2002; Molinie et al., 2005; Nguyen et al., 2007; Juan et al., 2008). This is particularly important in regard to the possible synergistic and/or additive effects of these mycotoxins that produce adverse toxicological effects to humans. The European Union (EU) Scientific Committee for Human Feeding has fixed the regulatory limits for OTA as 5 μg/kg in raw cereals (European Commission, 2005). Our results (0.5 to 8.25 μg/kg) show that the level of OTA alone as well as in combination with CTN can exceed the highest permissible limit in EU. Since only very limited toxicological data for CTN is available, CTN has not yet been brought into the purview of EU regulations. No maximum limit has been prescribed for

CTN in food by the Food and Agriculture Organization (FAO) (FAO, 2004) and there is no Indian regulatory standards prescribed for OTA and CTN alone as well as in combination in food products (PFA, 2008). Hence, there is a need for more data regarding the occurrence of CTN in food to set the maximum permissible limit.

Rice, corn and groundnut are the most staple food, coarse grain and most used oil seed, respectively. Hence, the average consumption of these food commodities has been estimated as distributed among the different States of India (FAS, 2011). Provisional Tolerable Weekly Intake (PTWI) of OTA is estimated as 100 ng/kg bw/week by Joint FAO/WHO Expert Committee on Food Additives (JECFA) and the Tolerable Daily Intake (TDI) estimated by EU legislation for OTA is 5 ng/Kg bw/day. Based on the data on consumption in Tamil Nadu and the average PTWI and TDI values, the result shows that OTA exceeded four times the EU recommendation and 1.4-fold higher than JECFA recommendation. Our results match the pervious findings on contaminated rice collected from Vietnam (Nguyen et al., 2007). However, in corn and groundnut samples the contamination of OTA was not exceeding the regulatory limits. So far, there is no regulatory limit of PTWI and TDI available for CTN. Therefore, Indian regulatory authority has to set the tolerable limit for CTN. It is important to consider the OTA and CTN co-contaminated food commodities because at lesser concentrations chronic exposure of OTA and CTN in combination would produce adverse health effects that are more serious than the individual exposures (Gayathri et al., 2015).

Toxigenic fungi that belong to the genera *Aspergillus* and *Penicillium* cause mycotoxin contamination of stored food commodities. Therefore, we were guided to isolate the fungi from the samples of rice, corn and groundnut. The results show that the incidence of *Aspergillus* sp was more than *Penicillium* sp. This is not surprising because the tropical climate of Tamil Nadu, where temperature is 24-33°C, promotes the growth of more the *Aspergillus* sp than *Penicillium* sp (Zimmerli and Dick, 1996; Bhat and Vasanthi, 2003). In order to pre-screen the mycotoxin-producing fungi belonging to *Aspergillus* and *Penicillium*, CCA plating method was adopted, and occurrence of mycotoxin-producing fungi was confirmed. From the results it is seen that the distribution of mycotoxin-producing *Penicillium* sp was less in rice, corn and groundnut samples. This is consistent with pervious observation on fungal contamination of food (Mills et al., 1989). The basis of the less abundant distribution of mycotoxin-producing *Penicillum* is not clear. Besides, the very presence of mycotoxigenic fungi in foodstuffs does not indicate mycotoxin contamination, but provides information on the potential risk of mycotoxin production (Bragulat et al., 2008). The crude extract of all the 15 highly fluorescing fungal isolates, identified from CCA, were further checked for the production of OTA and CTN using HPLC, and the results revealed that only *Aspergillus* fungal isolates produced either OTA or CTN. None of the *Penicillium* isolates produced OTA or CTN. Importantly, none of the single fungal isolates isolated from rice, corn and groundnut produced both the mycotoxins concurrently. The conditions underlying the production of more than one mycotoxin by the contaminating fungi may differ based on the interaction among the fungi, the nature of the substrate and the climatic conditions. Hence, the appropriate combination of these factors determines the mycoflora that infest the substrates, viz., rice, corn and groundnut, and the mycotoxin production (Moreno et al., 2009).

Based on the molecular characterization we could find *A. tubingensis* in rice and corn and *A. niger* in groundnut as OTA-producers. Previous reports have attested OTA production by *A. tubingensis* and *A. niger* (Pitt, 1987; Abraca et al., 1994; Wolff, 2000). CTN producers, *A. flavus, A. niger,* and *A. oryzae* were identified from rice, corn and groundnut. However, no special investigation and identification of CTN producers in rice, corn and groundnut has so far been attempted in India. This is first time it is reported that *A. flavus*, *A. niger* and *A. oryzae* produce CTN in tropical regions like Tamil Nadu in the peninsular India. Further work is necessary to address the climatic effect and quantitative analysis of mycotoxin synthesis by these OTA- and CTN- producing *Aspergillus* sp.

OTA and CTN co-occurrence was noticed after the outbreak of Balkan endemic nephropathy in Bulgarian villages (Petkova-Bocharova et al., 1990; Vrabcheva et al., 2000). The concentration of OTA and CTN found in cereals varies between <0.5 and 140 mg/kg and <5 and 420 mg/kg, respectively (Vrabcheva et al., 2000). The average concentration of OTA and CTN in rice, corn and groundnut samples collected from Tiruchirappalli district of Tamil Nadu, as seen in our results, also indicates that prolonged exposure of OTA and CTN alone as well as in combination would possibly cause health risk related to kidney disorders to Indian population. Recent studies indicate that 7 million patients in India are suffering from chronic kidney problems and, out of them, the etiological factor for 400,000 cases (5.4%) is unknown (Dash and Agarwal, 2006). Therefore, chronic exposure to OTA and CTN alone as well as in combination, by consuming contaminated rice, corn and/or groundnut, could be one of the unknown factors responsible for kidney problems in Indian population which needs further elucidation.

5. Conclusion

In spite of the limited number of samples examined, the data obtained from this research confirm the occurrence of OTA or CTN alone or co-contamination of OTA+CTN could be present in the rice, corn and groundnut samples collected from Tiruchirappalli district of Tamil Nadu. Since the co-contamination of OTA and CTN is toxic to kidney and liver at low concentrations it is a challenge to human and animal health. Thus, OTA and CTN, separately as well as in combination, pose health risk to humans and animals, which is an aspect of medical/clinical importance. The cause of the presence of mycotoxins in the stored food grains and cereals is the toxigenic fungal infestation. Therefore, the storage practices have to be improved to avoid the fungal contamination. Rural populations should be educated / sensitized about the possibility of multiple mycotoxin contamination, and all the food products that are destined for human and/or animal consumption have to be checked routinely for mycotoxin contamination to avoid the health risk.

ACKNOWLEDGMENT

The financial assistance from Doerenkamp-Zbinden Foundation, Switzerland, is heartily acknowledged.

REFERENCES

Abarca, M.L., Bragulat, M.R., Castella, G. and Cabanes, F.J. (1994). Ochratoxin A production by strains of *Aspergillus niger* var. *niger*. *Applied & Environmental Microbiology*, 60(7): 2650-2652.

Barah, B. (2009). Economic and ecological benefits of System of Rice Intensification (SRI) in Tamil Nadu. *Agricultural Economics Research Review*, 5(9): 209-214.

Bhat, R.V., Vasanthi, S., Rao, B.S., Rao, R.N., Rao, V.S., Nagaraja, K.V., Bai, R.G., Prasad, C.K., Vanchinathan, S., Roy, R. and Saha, S. (1997). Aflatoxin B1 contamination in maize samples collected from different geographical regions of India- a multicentre study. *Food Additives & Contaminants*, 14(2): 151-156.

Bhat, R.V., Vasanthi, S., Rao, B.S., Rao, R.N., Rao, V.S., Nagaraja, K.V., Bai, R.G., Prasad, C.K., Vanchinathan, S., Roy, R. and Saha, S. (1996). Aflatoxin B1 contamination in groundnut samples collected from different geographical regions of India: a multicentre study. *Food Additives & Contaminants*, 14(2):325-331.

Bouslimi, A., Bouaziz, C., Ayed-Boussema, I., Hassen, W. and Bacha, H. (2008). Individual and combined effects of ochratoxin A and citrininon viability and DNA fragmentation in cultured Vero cells and on chromosome aberrations in mice bone marrow cells. *Toxicology*, 251(1-3):1-7.

Bragulat, M.R., Martínez, E., Castellá, G. and Cabañes, F.J. (2008). Ochratoxin A and citrinin producing species of the genus Penicillium from feedstuffs. *International Journal of Food Microbiology*, 126(1-2): 43-48.

Dash, S.C., and Agarwal, S.K. (2005). Incidence of chronic kidney disease in India. *Nephrology Dialysis Transplantation*, 21(1):232-233.

Dyer, S.K., and McCammon, S. (1994). Detection of toxigenic isolates of *Aspergillus flavus* and related species on coconut cream agar. *Journal of Applied Bacteriology*, 76(1): 75-78.

El-Maghraby, O.M.O., and El-Maraghy, S.M. (1987). Mycoflora and mycotoxins of peanut (*Arachishypogaea* L.) seeds in Egypt. 1-sugar fungi and natural occurrence of mycotoxins. *Mycopathologia*, 98(3): 165-170.

European Commission (2005). Commission regulation (EC) No 123/2005 of 26 January 2005 amending regulation (EC) No 466/2001 as regards ochratoxin A. *Official Journal of the European Union* 25: 3-5.

FAO (Food and Agriculture Organization) (2004). Worldwide regulations for mycotoxins in food and feed. *FAO Food and Nutrition Paper*: 81.

Föllmann, W., Lebrun, S., Kullik, B., Koch, M., Römer, H.C. and Golka, K. (2000). Cytotoxicity of ochratoxin A and citrinin in different cell types *in vitro*. *Mycotoxin Research*, 16(1):123-126.

Kumar, R., Ansari, K.M., Saxena, N., Dwivedi, P.D., Jain, S.K. and Das, M. (2012). Detection of ochratoxinA in wheat samples in different regions of India. *Food Control*, 26(1): 63-67.

Gautam, A.K., Gupta, H. and Soni, Y. (2012). Screening of fungi and mycotoxins associated with stored rice grains in Himachal Pradesh. *International Journal of Theoretical and Applied Science*, 4(2): 128-133.

Gayathri, L., Dhivya, R., Dhanasekaran, D., Periasamy, V.S., Alshatwi, A.A. and Akbarsha, M.A. (2015). Hepatotoxic effect of ochratoxin A and citrinin, alone and in combination, and protective effect of vitamin E: In vitro study in HepG2 cell. *Food and Chemical Toxicology*, 83: 151-163

Karthikeyan, V., Patharajan, S., Palani, P. and Spadaro, D. (2010). Modified simple protocol for efficient fungal DNA extraction highly suitable for PCR based molecular methods. *Global Journal of Molecular Sciences*, 5(1): 37-42.

Heenan, C.N., Shaw, K.J. and Pitt, J.I. (1998). Ochratoxin A production by *Aspergillus carbonarius* and *A. niger* isolates and detection using coconut cream agar.*Journal of Food Mycology*, 1(2): 67-72.

IARC working group on the evaluation of carcinogenic risks to humans. (2007). 1,3-Butadiene, ethylene oxide and vinyl halides (vinyl fluoride, vinyl chloride and vinyl bromide) vinyl chloride. *IARC Monographs on the Evaluation of Carcinogenic Risks to Humans*, vol. 97: 311–443.

Jackson, L.K. and Ciegler, A. (1978). Production and analysis of citrinin in corn. *Applied Environmental Microbiology,*36(3): 408-411.

Janardhana, G.R., Raveesha, K.A. and Shetty, H.S. (1999). Mycotoxin contamination of maize grains grown in Karnataka (India). *Food and Chemical Toxicology*, 37(8): 863-868.

Jelinek, C.F., Pohland, A.E. and Wood, G.E. (1989). Worldwide occurrence of mycotoxins in foods and feeds--an update. *Journal-Association of Official Analytical Chemists*, 72(2): 223-230.

Juan, C., Zinedine, A., Idrissi, L. and Mañes, J. (2008). Ochratoxin A in rice on the Moroccan retail market. *International Journal of Food Microbiology*, 126(1): 83-85.

Kabak, B. (2009). The fate of mycotoxins during thermal food processing. *Journal of the Science of Food and Agriculture*, 89(4): 549-554.

Kishore, G.K., Pande, S., Manjula, K., Rao, J.N. and Thomas, D. (2002). Occurrence of mycotoxins and toxigenic fungi in groundnut (*Arachishypogaea* L.) seeds in Andhra Pradesh, India. *The Plant Pathology Journal*, 18(4): 204-209.

Klarić, M., Rašić, D. and Peraica, M. (2013). Deleterious effects of mycotoxin combinations involving ochratoxin A. *Toxins*, 5(11): 1965-1987.

Kumar, V., Basu, M.S. and Rajendran, T.P. (2008). Mycotoxin research and mycoflora in some commercially important agricultural commodities. *Crop Protection*, 27(6): 891-905.

Mills, J.T., Abramson, D., Frohlich, A.A. and Marquardt, R.R. (1989). Citrinin and ochratoxinA production by *Penicillium* spp. from stored durum wheat. *Canadian Journal of Plant Pathology*, 11(4): 357-360.

Molinié, A., Faucet, V., Castegnaro, M. and Pfohl-Leszkowicz, A. (2005). Analysis of some breakfast cereals on the French market for their contents of ochratoxin A, citrinin and fumonisin B1: development of a method for simultaneous extraction of ochratoxin A and citrinin. *Food Chemistry*, 92(3): 391-400.

Moreno, E.C., Garcia, G.T., Ono, M.A., Vizoni, É., Kawamura, O., Hirooka, E.Y. and Ono, E.Y.S. (2009). Co-occurrence of mycotoxins in corn samples from the Northern region of Paraná State, Brazil. *Food Chemistry*, 116(1): 220-226.

Nguyen, M.T., Tozlovanu, M., Tran, T.L. and Pfohl-Leszkowicz, A. (2007). Occurrence of aflatoxin B1, citrinin and ochratoxinA in rice in five provinces of the central region of Vietnam. *Food Chemistry*, 105(1): 42-47.

Osborne, B.G., Ibe, F., Brown, G.L., Petagine, F., Scudamore, K.A., Banks, J.N., Hetmanski, M.T. and Leonard, C.T. (1996). The effects of milling and processing on wheat contaminated with ochratoxin A. *Food Additives & Contaminants*, 13(2): 141-153.

Petkova-Bocharova, T., Castegnaro, M., Michelon, J. and Maru, V. (1991). Ochratoxin A and other mycotoxins in cereals from an area of Balkan endemic nephropathy and urinary tract tumours in Bulgaria. *IARC Scientific Publications*, 115: 83-87.

Pitt, J.I. (1987). Penicilliumviridicatum, Penicilliumverrucosum, and production of ochratoxin A. *Applied Environmental Microbiology,*53(2): 266-269.

Pitt, J.I. and Hocking, A.D. (2009). *Fungi and Food Spoilage* 519. New York: Springer.

Puntarić, D., Bošnir, J., Šmit, Z., Škes, I. and Baklaić, Z. (2001). Ochratoxin A in corn and wheat: geographical association with endemic nephropathy. *Croatian Medical Journal*, 42:175-180.

Ramesh, J., Sarathchandra, G. and Sureshkumar, V. (2013). A validated HPTLC method for detection of Ochratoxin A and citrinin contamination in feed, fodder and ingredient samples. *International Journal of Current Microbiology and Applied Science*, 2: 350-356.

Reddy, K.R.N., Salleh, B., Saad, B., Abbas, H.K., Abel, C.A. and Shier, W.T. (2010). An overview of mycotoxin contamination in foods and its implications for human health. *Toxin Reviews*, 29(1): 3-26.

Scott, P.M. (2005). Biomarkers of human exposure to ochratoxin A. *Food Additives and Contaminants,* 22(s1): 99-107.

Serra, R., Abrunhosa, L., Kozakiewicz, Z. and Venâncio, A., 2003. Black Aspergillus species as ochratoxinA producers in Portuguese wine grapes. *International Journal of Food Microbiology*, 88(1): 63-68.

Streit, E., Schatzmayr, G., Tassis, P., Tzika, E., Marin, D., Taranu, I., Tabuc, C., Nicolau, A., Aprodu, I., Puel, O. and Oswald, I.P. (2012). Current situation of mycotoxin contamination and co-occurrence in animal feed—Focus on Europe. *Toxins*, 4(10): 788-809.

Tamura, K., Stecher, G., Peterson, D., Filipski, A. and Kumar, S. (2013). MEGA6: molecular evolutionary genetics analysis version 6.0. *Molecular Biology and Evolution*, 30(12): 2725-2729.

Thirumala-Devi, K., Mayo, M.A., Reddy, G. and Reddy, D.V.R. (2002). Occurrence of aflatoxins and ochratoxinA in Indian poultry feeds. *Journal of Food Protection*, 65(8):1338-1340.

Vrabcheva, T., Usleber, E., Dietrich, R. and Märtlbauer, E. (2000). Co-occurrence of ochratoxin A and citrinin in cereals from Bulgarian villages with a history of Balkan endemic nephropathy. *Journal of Agricultural and Food Chemistry*, 48(6): 2483-2488.

Wagacha, J.M. and Muthomi, J.W. (2008). Mycotoxin problem in Africa: current status, implications to food safety and health and possible management strategies. *International Journal of Food Microbiology*, 124(1): 1-12.

White, T.J., Bruns, T.D., Lee, S.B. and Taylor, J.W. (1990). Amplification and direct sequencing of fungal ribosomal RNA genes for phylogenetics. In: Innis, M.A., Gelfand, D.H., Sninsky, J.J. and White, T.J., Eds., *PCR Protocols: A Guide to Methods and Applications*, Academic Press, New York, 315-322.

Wolff, J. (2000). Ochratoxin A in cereals and cereal products.*Archivfür Lebensmittel hygiene,* 51(4/5): 85-88.

World Health Organization (WHO) (1999). *High-Dose Irradiation: Wholesomeness of food irradiatied with doses above 10 kGy. World Health Organization*, 890: 1-149.

Zimmerli, B. and Dick, R. (1996). Ochratoxin A in table wine and grape-juice: occurrence and risk assessment. *Food Additives & Contaminants*, 13(6): 655-668.

In: Ochratoxin A and Aflatoxin B1
Editor: Reuben Hess

ISBN: 978-1-53617-416-8

Chapter 4

PRESENCE OF AFLATOXIN B1 IN BEER: CONTAMINATION DURING PROCESSING AND METHODS OF ANALYSIS

Antonio Ruiz-Medina*
Department of Physical and Analytical Chemistry,
University of Jaén, Jaén, Spain

ABSTRACT

Besides the known adverse effects of alcohol, beer could also be the source of several mycotoxins such as aflatoxin B1 (AFB1), the most toxic one among the identified aflatoxins and classified into group I as a human carcinogen by the International Agency for Research on Cancers. For this reason, the contamination of cereals and therefore beer is controlled worldwide by legal limits to ensure public health safety.

The appearance of AFB1 in beer is the result of contamination from malted grain or food additives. The fate and stability of AFB1 during brewing process (malting, mashing, fermentation, maturation, etc.) as well as the most important detection techniques and sample treatment used for the determination of this compound will be evaluated in this chapter.

---

* Corresponding Author's Email: anruiz@ujaen.es.

Keywords: aflatoxin B1, beer, brewing, transfer, analytical methods, decontamination

INTRODUCTION

Mycotoxins are toxic secondary metabolic products of molds present on almost all agricultural commodities worldwide. Currently there are around 400 mycotoxins reported. These compounds occur under natural conditions in feed as well as in food, especially cereals and cereal products. Mycotoxins are produced by different strains of fungi and each strain can produce more than one. The major classes of these mycotoxin-producing fungi are *Aspergillus, Penicillium* and *Fusarium*. Mycotoxins are invisible, tasteless, chemically stable and resistant to temperature and storage. They are resistant to the normal feed manufacturing processes and can cause economic losses at all levels of production, including crop and animal production, processing and distribution. According to the FAO (Food and Agriculture Organization) around 50% of the world's crop harvests may be contaminated with mycotoxins.

The main toxic effects of mycotoxins are carcinogenicity, genotoxicity, nephrotoxicity, hepatotoxicity, estrogenicity, reproductive and digestive disorders, immunosuppression and dermal effects. The maximum levels and codes of practice the prevention and reduction have been instituted by the Codex Alimentarius Commission in order to minimize mycotoxin levels in various foods. These regulations have been adopted for foods in many countries; the European Union (EU), especially, places strict regulations on many different kinds of mycotoxins in a wide range of foods.

Some of the most common mycotoxins include: aflatoxins (AFs), trichothecenes, fumonisins, zearalenone, ochratoxin A and ergot alkaloids. Nevertheless, the best known and most intensively researched mycotoxins in the world are probably AFs which are produced by different species of *Aspergillus* (mainly *Flavus*, *Parasitous* and *Nomius*), a fungus which is especially found in areas with hot and humid climates. The factors implicated in the growth of this fungus in foods are as much those relating to the environment in which they develop (pH, composition of the food or water activity) as to extrinsic factors: ambient humidity, storage temperature and microbial competition. They can occur in foods such as groundnuts, treenuts, maize, corn, rice, figs and other dried foods, spices, crude vegetable oils and cocoa beans, as a result of fungal contamination before and after harvest during storage if water is allowed to exceed critical values for mould growth.

AFs were discovered at the end of the 1950s and beginning of the 1960s as a result of an investigation conducted to evaluate the high mortality rate in poultry and other food-producing animals as a consequence of ingesting feed containing peanuts originating from South America. The ability of AFs to cause cancer and related diseases in humans given their seemingly unavoidable occurrence in food and feed makes the prevention and detoxification of these mycotoxins one of the most challenging toxicology issues of today. Exposure through food should be kept as low as possible. The products specified in the regulations which contain AFs in levels exceeding the established maximum content must not be marketed as such, even after being combined with other food products, and must not be used as an ingredient in other foods. Among 18 different types of AFs identified, major members are AFs B1 (AFB1), B2, G1, G2, M1 and M2. AFB1 is the most common in food and the most potent natural carcinogen known. AF M1 is its major metabolite in humans and animals.

Beer is a very old drink in history in which the making of ferment beverages was discovered by primitive humans and has been practiced as an art for thousands of years. This drink is one of the products that is susceptible to mycotoxin contamination. Cereals used in beer production, particularly barley, wheat and maize, can be contaminated by different mycotoxins. The contamination can occur both in field and during storage, if conditions of high moisture are present, and during malting. The raw material that is capable of supporting microbial growth has the potential to produce unwanted metabolites that can persist through the brewing process and produce adverse effects. Since cereals are raw materials of beer and beer-based drinks, it is necessary to consider that the risk of contamination exists in these products.

The most abundant mycotoxin in beer is found to be deoxynivalenol. However, other studies show the presence of toxins such as zearalenone, AFs, fumonisins B1, B2 and B3, ochratoxin A together with their modified forms. The EU Regulation EC 1881/2006 establishes maximum allowed levels for mycotoxins, however the modified forms are not yet included. The limits for cereal based products, such as beer, are set as follows: 2 μg kg^{-1} for AFB1 and 4 μg kg^{-1} for total AFs, 750 μg kg^{-1} for deoxynivalenol, 75 μg kg^{-1} for zearalenone, 400 μg kg^{-1} for total fumonisins and 5 μg kg^{-1} for ochratoxin A. The majority of modified mycotoxins are less toxic than their parent forms, nevertheless, enzymes present in the digestive system may be able to transform the modified forms into parent forms and may have consequences on human health that are difficult to predict. Reducing the fungal contamination of malt and barley is the most promising strategy for reducing AFB1 in beer. At present, the EU has not set a maximum allowable limit for AFB1 in beer. Beer's preparation process and transfer of AFB1 in this matrix will be evaluated. Moreover, different techniques used for the determination of this compound will be also studied along this chapter.

Brewing Process

Brewing is the process of production of malt beverages, specifically, it is the practice of regulating the interactions between water, starch, yeast, and hops so that the end result is what is called "regular beer." The yeast turns sugars in the malt into alcohol and the hops provide the bitter flavor in beer and the flowery aroma. The flavor of the beer depends on many things, including the types of malt and hops used, other ingredients and the yeast variety. The art of brewing is as old as civilization, although beer production and consumption is booming like never before, mainly due to the increasing popularity of craft beer. Craft beer is produced by small, independent and traditional breweries according to the definition of the Brewers Association. The number of craft breweries continues to grow, claiming a larger market share every year. The reason for the popularity of craft brewers is that they tend to focus on flavour and tradition, combined with innovation rather than on large-scale and low-cost production. Whereas "regular beer" is brewed almost exclusively with the compounds previously mentioned, craft brewers add a wide range of different ingredients to the brewing process. Some examples are coffee, cacao, tobacco, liquorice, nuts, tomatoes, chili peppers, fruit and a range of spices (Oliver, 2011).

Getting the yeast right is essential as each variety has its own distinctive effect on the beer. The basic process may be simple but the execution is highly sophisticated (Figure 1). The brewing process to obtain "regular beer" consists of eight key components: malting; milling; mashing; boiling of wort; cooling; fermentation; maturation, carbonation and filtering; and finally, packaging. A description of each of these steps will follow.

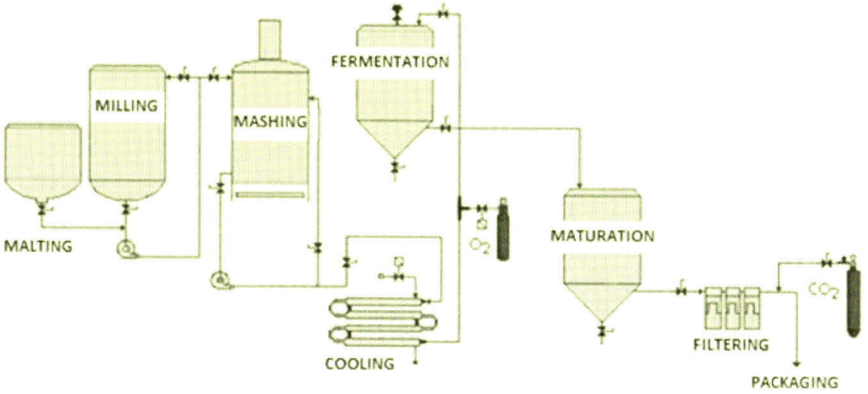

Figure 1. The brewery process.

1. *Malting*. It is the process of readying barley to be used in brewing. Firstly, the grain is steeped (soaked) in a vat of water for about 40 hours and next germination of the barley grain is produced. At the end of this process, about three days, the starch has become soft and the enzymes have not started converting the starch into sugar yet. The barley grain is called green malt. This step allows the formation of highly active α-amylase, β-amylase and proteases enzymes as well as various flavor and color components. Secondly, germination is halted by drying the green malt on metal racks in the kiln house

(slow heating at 80°C for a light malt, or higher for a dark malt). The kilning temperature must not harm amylase enzyme. Furthermore, if kilning temperature is higher, darker will be the beer produced. Although malted barley is the primary ingredient, unmalted corn, rice or wheat are sometimes added, to produce different beer flavours.

2. *Milling.* It is the cracking of the grain between rollers to produce coarse powder, the grist, which the brewer chooses for the particular batch of beer. This process need to be manipulated carefully to find a balance between a grind that is too fine and one that is too coarse. The finer the grind, the greater will be the amount of sugary wort that can be extracted from a given amount of grist. Milling the grain allows it to absorb the water it will eventually be mixed with in order for the water to extract sugars from the malt.
3. *Mashing.* It is the process of turning the grist into a sweetened liquid. Mashing converts the starches, which were released during the malting stage, to sugars that can be fermented. The milled grain is dropped into warm water then gradually heated to around 60-75°C in a large cooking vessel called the mash tun. In this mash tun, the grain and heated water mix creating a cereal mash to dissolve the starch into the water, transforming it into sugar (mainly maltose). The degree of enzymatic hydrolysis is strongly depends on pH and temperature. β-amylase has optimum activity at temperature 60-65°C whereas α-amylase has optimum activity at temperature 70-75°C. Because water is such a vital part of the brewing process, the water itself is a key ingredient. This sugar rich water is then strained through the bottom of the mash and is now called wort.
4. *Boiling of wort.* The spent grains and precipitated proteins are filtered out and the wort is ready for boiling which involves many technical and chemical reactions. Reasons for boiling the wort: extraction of hop flavour from hop flower, coagulation of remaining protein, inactivation of enzymes that were active during mashing and, finally, also sterilize and concentrate the wort. During this stage, important decisions will be made affecting the flavor, color

and aroma of the beer. Hop flower, dried female flower of hop plant *Humulus Lupulus*, are added at various intervals for either bitterness or aroma and to help preserve it. Approximately one quarter pound of hop flower is added per barrel of beer and up to 2 pound per barrel of ale. Normally, maize grit is used. The wort is boiled for one to two hours to sterilize and concentrate it, and extract the necessary essence from the hops. The pH is adjusted to 5.5 ± 0.5 by means of phosphoric acid.

5. *Cooling.* The wort is transferred quickly from the brew kettle through a device to filter out the hops, and then onto a heat exchanger to be cooled. The heat exchanger basically consists of tubing inside of a tub of cold water. It is important to cool the wort quickly before oxidation or contamination can occur. While the wort is still hot, bacteria and wild yeasts are inhibited. Moreover, there are also the sulfur compounds that evolve from the wort while it is hot (dimethyl sulfide will continue to be produced in the wort without being boiled off), causing off-flavors in the finished beer. Rapid cooling also has the benefit of causing more of the naturally occurring haze-proteins in the wort to settle in the kettle, so finished beer will be clearer. The hopped wort is saturated with air, essential for the growth of the yeast in the next stage.
6. *Fermentation.* After passing through the heat exchanger, the cooled wort goes to the fermentation tank. The brewer now selects a type of yeast and adds it to the fermentation tank in order to this one eats the sugar in the wort and turns it into ethanol and carbon dioxide plus some amount of glycerol and acetic acid. Most beer contains 3.5-5% alcohol. This process of fermentation takes 10-14 days, approximately. Although this process is usually carried out at 3-4°C, it may range from 3-14°C. The wort finally becomes beer. Each brewery has its own strains of yeast, and it is these that largely determine the character of the beer. In some yeast varieties, the cells rise to the top at the end of fermentation and are then skimmed off. This is called top fermentation, and ales are brewed in this way. When at the end of fermentation the yeast cells sink to the bottom,

the process is known as bottom fermentation, used for lager or pils. Open tank fermenter can be used, however, closed fermenter tank is preferred so that CO_2 liberated during fermentation can be collected for later carbonation step. CO_2 evolution is maximum by fifth day of fermentation; there is no evolution from the 7-9 days because yeast cells become inactive and flocculate.

7. *Maturation, carbonation and filtering.* The beer has now been brewed, but it can still be improved through maturation (from weeks to several months). During this phase, the brewer moves, or racks, the beer into a new tank called the conditioning tank. The brewer then waits for the beer to complete its aging process, occurring the precipitation of proteins, yeast, resin and other undesirable substances. The taste ripens. The liquid clarifies and it is filtered and carbonated by 0.45-0.52% CO_2. Further filtering gives the beer a sparkling clarity (diatomeaus earth and polyvinyl polypyrrolidine can be used as the main filter aids).
8. *Packaging.* The beer is moved to a holding tank where it stays until it is bottled, pasteurized (kept at ~65ºC for 20 min and then cooled to room temperature), canned or put into kegs. Packaged beer is pasteurized in order to provide a shelf-life of up to six months when properly stored. Filling techniques ensure air does not come into contact with the beer, and cannot be trapped within the container. Draught beer, since it is normally sold and consumed within a few weeks, may not go through this process. The draught beer is placed in sterilized kegs, ready for shipment.

AFB1 Transfer to Beer

Since beer plays an important role in the human diet, and any foodstuff consumed in such large quantities is a potential path for ingestion of harmful substances, it is very important to control the presence of mycotoxins in this commodity. Studies with different mycotoxins added at various stages of the brewing process show that AFB1 may be transmitted from contaminated grains into beer. They originate from the malted grain or from food additives. The natural occurrence in hops of AFB1, maize for example, has been reported in various studies; so, as a consequence of the use of these adjuncts in the brewing process, AFB1 may potentially occurs in beer. The first experiments to trace the AFB1 fate in the brewing process were carried out in the 1970s when the authors (Chu et al., 1975) added before the malting process purified standard of AFB1 at concentrations of 1 and 10 mg g^{-1} of malt. A great portion of AFB1 was partially lost during the brewing. It was relatively stable in the cooker mash step but more sensitive to later treatments, especially the wort boiling and final fermentation. Since AFB1 is relatively stable to heat, it is not surprising that more than 90% of this one was in the sample after cooker mash treatment. The significant loss of AFB1 between the combined mash and boiled wort steps may be due to the non-specific interaction or adsorption of AFB1 by the solid particles removed by the filtration process.

Trace amounts of AFB1 were found in several samples of beers imported from south Europe in the 1990s. They occurred in beers coming from South America or equatorial Asian countries, at levels lower than 0.09 mg l^{-1} (Scott & Lawrence, 1996). AFs (B1, B2, G1 and G2), zearalenone, citrinin, deoxynivalenol, and ochratoxin A were studied in traditionally brewed beers consumed by most ethnic black South Africans (Odhav & Naicker, 2002). No mycotoxin-producing fungi were present in the fermented beers but two of six commercial beer samples contained AFs (200 and 400 μ gl^{-1}) and 45% (13 of 29) of the home-brewed beers had zearalenone (range 2.6-426 μg l^{-1}) and/or ochratoxin A (3-2340 μg l^{-1}). In a following survey on 304 samples of Canadian and imported beer from 36 countries, the toxin was found only in some samples from Mexico, Spain and Portugal (Mably et al., 2005).

The transfer of AFB1 from naturally contaminated raw materials to beer during an industrial brewing process was also evaluated in another study (Pietri et al., 2010). For each stage of the brewing, the AFB1 percentage in the product, referring to the total amount in the raw materials (TARM), was calculated. A relatively small amount of the toxin was transferred to the liquid phase during mashing; only 14% of the AFB1 TARM was found in the liquor after mashing. In the sweet wort the AFB1 level was reduced both because of the dilution of the product, and mainly because of the filtration step: only an average of 8% of the AFB1 TARM was found in the sweet wort. In the following steps, the reductions were more moderate (boiling -28%, cooling -3%, fermentation -14%), while a further strong decrease (-60%) was observed passing from green to bottled beer. The average percentage of AFB1 recovered in finished beer, referring to the amounts contained in raw materials, was 1.5% ± 0.8%. This result was mainly due to its solubility during the mashing process. If raw materials comply with the limits fixed by EU regulations, 2 μg Kg^{-1}, the contribution of a moderate daily consumption of beer to AFB1 intake does not contribute significantly to the exposure of the consumer.

In 2008-2011 a total set of 333 samples of brewing raw materials and beer were analyzed (Beneŝová et al., 2012) for the presence of AFB1 and AFs B2, G1 and G2. In 7 of 216 samples of brewing raw materials (3.2%), AFs were found at trace concentrations to 1.2 mg kg^{-1}. On the other hand, in 6 of 117 (5.1%) beer samples, AFs were detected at concentrations to 31 ng l^{-1} (values exceeding the quantitation limit (LOQ)). The occurrence of AFB1 and other mycotoxins in a sample of 106 beers produced in several European countries was also investigated (Bertuzzi et al., 2011). AFB1 was not detected in any sample, whereas ochratoxin A, deoxynivalenol and fumonisins were found in a relatively high number of samples. A study for the selective determination of AFB1 in different types of beer was recently developed (Molina-García et al., 2012). Of the 17 beers analysed, including normal (n = 12) and non-alcoholic (n = 5) beers, only three samples contained traces of the target compound (0.86, 0.28 and 0.17 mg l^{-1}). AFB1 was not detected in the remaining beer samples investigated, including non-alcoholic beers. A total of 101 samples of beer from the Chinese market were analysed for the presence of AFB1 and sterigmatocystin (Zhao et al., 2017). None of the beer purchased samples were contaminated with both toxins. The LOQ and the detection limit (LOD) were 0.1 and 0.03 μg kg^{-1}, respectively, with recoveries from spiked beer samples between 97.8-103.6%.

The content of AFB1 in 384 samples from 16 trademarks of clear lager beers, produced by two Mexican breweries, was evaluated by Álvarez-Segovia et al., 2019, by using a competitive ELISA assay. Results showed that all beers contained this toxin in a range of 0.203–0.241 $\mu g\ l^{-1}$. Although Maximum Residue Levels (MRL) have not been established for AFB1 in beer, there is a limit for processed cereals; accordingly, none of the samples exceeded the MRL of AFB1 set by Mexican standard (20 $\mu g\ l^{-1}$) or by the EC for cereals and cereal-based products. In another research developed by Peters et al., 2017, the presence of mycotoxins in beers collected from 47 countries (60% craft beers) was studied. A selection of 1000 samples were screened for AFB1, ochratoxin A, zearalenone, fumonisins, T-2 and HT-2 toxins, and deoxynivalenol using a mycotoxin 6-plex immunoassay. Many samples had small amounts of these compounds although they did not pose any risk to human consumption.

AFs (B1, B2, G1 and G2) were surveyed in 417 beer samples purchased most of them from Spain in the period 2006-2012 (Burdaspal & Legarda, 2013), being detected in 72.6% of the samples with levels ranging from 0.08 to 36.12 ng l^{-1} for all four AFs. The median values obtained in this study for AFs in beers would result in an intake of approx. 97-112 pg/per capita/day, which represents a very small fraction (approximately 0.5% in a worst-case scenario) of the estimated average exposure to total AFs. The same AFs were studied in eighteen brands (15 domestic and 3 imported beers) collected from different market in Ethiopia (Nigussie et al., 2018). The result showed, out of twelve alcoholic domestic beer brands eleven were positive with a range of total AFs between 1.23 and 12.47 $\mu g\ l^{-1}$ and one brand was less than LOD. Finally, AFB1 was also evaluated in beer produced in sub-Saharan Africa (Lulamba et al., 2019). Residual levels of <20% can be achieved during the production. This is because the prevailing environmental conditions during beer production are favorable to mycotoxigenic fungal proliferation. This subsequently leads to relatively high concentration of AFB1 in freshly processed beer, with a possible increase during the beer shelf-life owing to the absence of appropriate microbial stabilization treatments in the finished processed beer.

Analysis Methodologies

The analytical determination of AFB1 is complicated by two main factors: (1) the complexity of the sample matrix in which it normally appears; and (2) the low levels present in the sample. Analytical methodologies must be designed to address these requirements. Firstly, to avoid matrix interferences, a typical analysis of AFB1 requires a treatment of the beer. Secondly, with respect to its detection, a reliable and sensitive method must be selected for the screening and determination of this compound.

Sample Treatment

In beer, there are a wide range of low molecular weight compounds such as sugars, pigments or organic acids, which cause significant matrix effect in the detection step. These interfering compounds are the main problem faced by multi-toxin methods since they may cause underestimation of the amount of toxin in samples. In order to minimize matrix interactions, many of multi-toxin methods applied in beer employ a sample dilution. Nevertheless, even after the dilution step, matrix-matched calibration or labeled internal standards are usually employed to correct the matrix effect and thus to perform a correct quantification (Peters et al., 2017). For this reason, most of the methods use a treatment of the sample. Tremendous efforts have been made to develop some novel, fast, efficient, economical, miniaturized and simplistic sample preparation methods. Some of them are discussed below.

Several multimycotoxin methods based on Solid Phase Extraction (SPE) procedure have been developed in different food commodities. This procedure involves a Liquid-Liquid Extraction (LLE) (using solvents such as methanol, acetonitrile and/or their combinations), followed by a clean-up step, e.g., multifunctional or immnunoaffinity columns. SPE with immunoaffinity material is very popular in mycotoxin analysis because it is a very selective and time saving sample cleanup tool (Visconti et al., 2000) for removing matrix compounds. However, one of the problems of the immunoaffinity materials is the high cost, and other alternatives have been checked with conventional sorbents such as C_{18}, hydrophilic-lipophilic balanced copolymers or ion exchangers (Romero-González et al., 2009). Molecularly Imprinted Polymers (MIPs) have also been used as SPE sorbents. They are tailor-made polymers with a predetermined selectivity toward a given analyte or a group of structurally related species. MIPs are prepared by the polymerization of suitable functional monomers and cross-linking agents in the presence of a molecular template. After polymerization, the template is removed from the polymeric matrix leaving cavities complementary in size and shape to the template. MIPs are a good election for elimination of AFs from grains, foods and feed samples (Wei et al., 2015).

Another alternative is to use the QuEChERS (Quick, Easy, Cheap, Effective, Rugged and Safe) method (Anastassiades et al., 2003). This methodology, initially developed for pesticide analysis, was examined in beer-based drinks (beer, low-malt beer, new genre, and nonalcoholic) (Tamura et al., 2011). This involves first an extraction step with acetonitrile-water followed by a liquid-liquid partitioning induced by the addition of inorganic salts (sodium chloride and anhydrous magnesium sulfate). As a result, some polar components of the matrix remain in the aqueous layer while mycotoxins are moved into the organic phase. Thereafter, a Dispersive Solid Phase Extraction (DSPE) is employed to reduce other matrix compounds from the organic phase. The extract is processed by shaking with either a primary-secondary amine (PSA) silica gel alone, or PSA plus C_{18} or graphite carbon black. In this way, the reduction of matrix-interfering compounds increases the sensitivity of the method.

The Dispersive Micro-Solid Phase Extraction (D-μ-SPE) process using zirconia nanoparticles as a dispersant per se is novel in terms of application for mycotoxin extraction. It has been introduced as a new version of SPE which holds the advantages of Dispersive Liquid-Liquid Extraction (DLLE) and SPE simultaneously. Compared with QuEChERS method, D-μ-SPE was found to be advantageous in terms of short time requirement, speed and reduced solvent, and sorbent consumption. The selection of the dispersant is a key point to achieve the appropriate selectivity, the satisfactory extraction efficiency and the high enrichment factor in the D-μ-SPE process. The conventional materials are being substituted by nanomaterials due to their high surface-to-volume ratio and high sorption properties (Du et al. 2018).

AFB1 Determination

As indicated before, AFB1 gets to beer either from contaminated input materials or adjuncts added in the course of brewing. For this purpose, reliable analytical methods for fast and effective monitoring of this compound in the beer production chain are needed at the low legislation established limit. Different methods have been established for the determination of AFB1 (Turner et al., 2009), including capillary electrophoresis and thin-layer chromatography (Chauhan et al., 2016), suffering however both from low sensitivity and difficulty for automation. Nevertheless, the two most important groups of methods applied to AFB1 determination are chromatographic ones (gas or liquid chromatography) and immunologic ones.

Table 1. Chromatographic methods for AFB1 determination

Method	Column	Sample Preparation	LOD	LOQ	Sample	Reference
LC-Fluorescence detection	-	Immunoaffinity	0.019 μg L^{-1}	-	Beer	Scott et al., 1997
LC-Fluorescence detection	HP SB-c18 (Chrom. Spec.)	Immunoaffinity	0.004 μg L^{-1}	0.0007 μg L^{-1}	Wine and beer	Mably et al., 2005
UHPLC-MS/MS	Acquity UPL (Waters)	SPE	0.1 μg L^{-1}	-	Beer	Ventura et al., 2006
LC-MS/MS	Pro Star 210 (Varian)	Immunoaffinity	-	-	Beer and sake	Rudrabhatla et al., 2008
UHPLC-MS/MS	Acquity UPL (Waters)	SPE	0.04 μg L^{-1}	0.13 μg L^{-1}	Beer	Romero et al., 2009

Method	Column	Sample Preparation	LOD	LOQ	Sample	Reference
UHPLC-orbitrapMS	Acquity HSS T3 (Waters)	Liquid-liquid extraction	-	0.5 μg L^{-1}	Beer	Zachariasova, et al., 2010
HPLC-Fluorescence detection	Superspher RP-18 (Merk)	Immunoaffinity	0.001 μg L^{-1}	0.003 μg L^{-1}	Beer	Bertuzzi et al., 2011
UHPLC-MS/MS	Acquity UPLC (Waters)	QuEChERS	-	1.0 μg L^{-1}	Beer-based drinks	Tamura et al., 2011
HPLC-MS	Synergi Hidro RP 80A	Immunoaffinity	0.0015 μg L^{-1}	0.005 μg L^{-1}	Brewing materials/beer	Benešová et al., 2012
LC-MS/MS	BEH C18	Liquid-liquid extraction	0.003 ng mL^{-1}	0.010 ng mL^{-1}	Nonalcoholic beer	Khan et al., 2013
LC-MS/MS	Symmetry C18	SPE	2.5 μg Kg^{-1}	5 μg Kg^{-1}	Beer	Matumba et al., 2014
HPLC-Fluorescence detection	TC C18	MIPs-SPE	0.05 μg Kg^{-1}	0.16 μg Kg^{-1}	Beer	Wei et al., 2015
LC-MS/MS	Gemini C18	SLE/QuEChERS	2 ng Kg^{-1}	3.5 ng Kg^{-1}	Barley and Beer	Juan et al., 2017
UHPLC-QTOF/MS	Eclipse plus C8 RRHD	Microwave-assisted D-μ-SPE	0.016 μg Kg^{-1}	0.049 μg Kg^{-1}	Beer	Du et al., 2018
LC-Fluorescence detection	Shimadzu CTO-20AC	Immunoaffinity	0.2 μg L^{-1}	0.8 μg L^{-1}	Beer	Nigussie et al. 2018
LC-MS/MS	Acquity UPLC (Waters)	Liquid-liquid extraction	3.22 μg L^{-1}	6.43 μg L^{-1}	Beer	Pascari et al., 2018
UPLC-MS/MS	ACQUITY HSS T3 (Waters)	QuEChERS/ D-μ-SPE	0.012 μg L^{-1}	0.038 μg L^{-1}	Beer	González-Jartín et al., 2019

LOD: Limit of Detection; LOQ: Limit of Quantitation. LC: Liquid Chromatography; UHPLC: Ultra High-Performance Liquid Chromatography; MS: Mass Spectrometry; MS/MS: Tandem Mass Spectrometry; HPLC: High-Performance Liquid Chromatography; QTOF: Quadrupole Time of Flight; UPLC: Ultra Performance Liquid Chromatography; SPE: Solid Phase Extraction; QuEChERS: Quick, Easy, Cheap, Effective, Rugged and Safe; MIPs: Molecularly Imprinted Polymers; SLE: Solid Liquid Extraction; D-μ-SPE: Dispersive Micro-Solid Phase Extraction.

The chromatographic methods measure the compound specifically after sample extraction and extract clean-up followed by a separation by Gas (GC) or Liquid Chromatography (LC) (Zollner & Mayer-Helm, 2006) and a specific detection. Nowadays, LC coupled to Mass Spectrometry (MS) or Tandem Mass Spectrometry (MS/MS) is mainly used for the simultaneous determination of several classes of mycotoxins because it provides more accurate identification, better selectivity, and higher sensitivity than other detection techniques although it requires expensive equipment. Furthermore, the introduction of Ultra Performance Liquid Chormatography (UPLC) hyphenated with MS/MS, high-resolution MS (Orbitrap MS) or time of flight (TOF)-MS technology offer fast analysis, improved peak resolution, low LODs and high sample throughput. The number of publications in this field has increased in the last years (Frenich et al., 2009), including applications to beer samples. Some data of different published methods are included in Table 1.

Despite the advantages of MS detection, the AOAC official method for the determination of AFB1 in foods and agriculture commodities is based on high-performance liquid chromatography (HPLC) with fluorescence detection (native fluorescence of AFB1) (AOAC Official method, 2000). Firstly, an extraction step with acetonitrile and deionized water followed by a clean-up procedure through a packing material are realized. The purified extract is transferred quantitatively from top of column to screw cap vial (derivatization vial) and is evaporated under nitrogen. Secondly, after adding n-hexane to re-dissolve AFB1, trifluoroacetic acid and acetonitrile are added to produce the AFs derivatization. Finally, fluorescence is measured by using 365 and 435 nm for excitation and emission wavelengths, respectively.

Immunochemical methods have emerged as alternatives for mycotoxin determination with high detection sensitivity at lower analysis cost. They offer increased potential for on-site determinations through appropriate miniaturization of the required instrumentation and development of rapid assays. Moreover, when a qualitative or semi-quantitative result is sufficient, lateral flow immunoassay strips could be applied. The main argument against the use of immunochemical versus the instrumental ones is the low level of multiplex capacity. Several immunosensors, based mainly on electrochemical or optical transduction principles, have been developed and applied for the detection of AFB1 in beer demonstrating excellent analytical performance (Pagkali et al., 2018; Beloglazova & Eremin, 2015). Kits to perform the Enzyme-Linked Immunosorbent Assays (ELISA) (Ghali et al., 2008) are available from different suppliers and these methods are often used for screening purposes and on the field. Different versions exist that vary in the rapidity of the procedure and the quantification of the results although all of them require enzymatic reactions, washing and separation of bound and free label. A very important step of validation is the detection of matrix effects, and since false positive can occur, positive results could require confirmation by more expensive chromatographic methods.

Sensors may be also a good choice for the analysis of AFB1 due to their fast, simple, and low-cost detection capabilities. Electrochemical sensors (Tan et al., 2009) can be found in literature for the determination of this AF but they have not been applied to alcoholic beverages. The use of spectrofluorimetric detection would be a possibility to develop an optical sensor but it is difficult due to the complexity of the matrix, which shows a great variety of natural fluorescent compounds whose spectra often overlap the analyte signal. As alternative to measure native fluorescence of AFB1, a method based on the on-line generation of a fluorescent photoproduct from this one by UV-irradiation and its monitoring when adsorbed onto C_{18} silica gel has been proposed (Molina-García et al., 2012). The schematic valve system diagram is shown in Figure 2. In the initial status, the UV lamp and the peristaltic pump were switched on to condition the flow system until a stable baseline was recorded. Being all valves switched off, the carrier solution (25% (v/v) methanol:water solution) flowed through the system, while the sample solution was recycled to its vessel. Then, in order to insert the sample in the flow system, valves V1, V2 and V3 were switched on: sample solution circulated through the system whereas the carrier solution was recycled to its recipient. Next, the flow was stopped so allowing the photodegradation process. Finally, when valve V1 was switched off and the AFB1 photodegradation product arrived to the flow-cell, the analytical signal was measured at the corresponding wavelengths (353/424 nm/nm for excitation and emission wavelengths, respectively). The measurement of the photoproduct retained on this solid support, packed in the flow-cell placed in the detection area, was adopted for improving both LOD and selectivity.

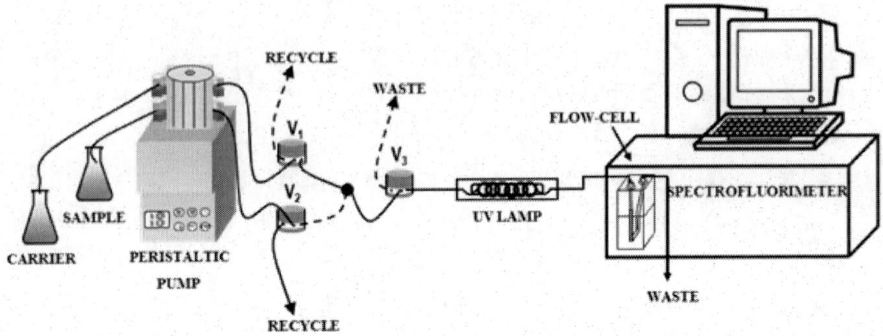

Figure 2. Flow-injection system for AFB1 determination (previous generation of its fluorescent photoproduct).

Alternatively, biosensors have emerged as an attractive and promising tool for diverse analytical applications (Lv et al., 2014; Xu et al., 2016). Despite their improved analytical merits, antibodies are considered a very expensive biorecognition element with the problem of long-term stability in addition to being produced by the animal source. One option in biosensors field is the use of aptamers for AFB1 detection. They are single-stranded nucleic acid or peptide molecules having high affinity and specificity to a wide range of targets, which can be produced by natural origin or synthetic procedure. Aptamers are mainly obtained by an in vitro selection process called SELEX (Systematic Evolution of the Ligand by the Exponential Enrichment Process). Aptamers offer inherent advantages with respect to antibodies such as chemical and thermal stability, ease of modification, in vitro production and low-cost synthesis. Moreover, aptamers are not susceptible to denaturation in the presence of solvents commonly used in the extraction of toxin analytes (Goud et al., 2016). Mostly the fluorescence quenching-based aptasensors are developed based on graphene quantum dots and nanoparticles. On the other hand, aptamer based surface plasmon resonance is an interesting alternative because small molecules such as AFB1 can cause little change of mass concentration at the biosensor surface (Sun et al., 2017). Finally, although electrochemical aptasensors require several steps and many calibration points, electrochemical techniques offers the possibility of portable and miniaturized devices and thus can be considered promising tools for on-site analysis of food contaminants.

Decontamination

Fungal infestation is almost inevitable due to weather conditions and the specificity of field treatment. Also, inappropriate barley storage conditions may take place because of equipment fail which could lead to microorganism activation and growth. Thus, in order to minimize economical losses, the need of fungal and mycotoxin decontamination becomes obvious (Pascari et al., 2018). Chemical treatment, such as ozonation, is of a promising future in barley and beer detoxification of mycotoxins (Piacentini et al., 2015), taking into account that this treatment do not leave any residual chemical.

On the other hand, lactic acid bacteria starter cultures added during malting and brewing could also represent an efficient strategy in mitigation of fungal and maybe mycotoxin contamination (Oliveira et al., 2015). The pH drop due to the presence of lactic acid is able to stimulate enzymatic activity during malting and fermentation. Besides, the use of microorganisms offers an attractive alternative for the control or elimination of AFs in foodstuffs, being *Saccharomyces cerevisiae* one of the most effective for binding AFB1 (Shetty et al., 2007). Products based on this microorganism (cell wall from baker and brewer yeasts, inactivated baker yeast or alcohol yeast) have been studied, showing that in pH 3, at 37ºC and 15 min of contact, the removal of AFB1 ranged from 2.5% to 49.3%, depending on the concentration of the toxin in the medium and on the yeast-based product used.

Although, as we have seen, there are different decontamination methods, the best way to preserve the food and the consumer is prevention. However, the aforementioned methods give a possibility to reduce food waste and stimulate a sustainable production.

CONCLUSION

Mycotoxins are potential problems for farmers and food producers because they can adversely affect production. Close attention should therefore be paid to the risk of contamination. According to World Health Organization (WHO), beer is the most consumed alcoholic beverage worldwide. Beers can be contaminated by mycotoxins due to the use of contaminated cereals for their production while some additional portion or other mycotoxins might arise during the beer brewing process raising significant concern for the possible effects on human health.

AFB1 is not completely removed from the raw materials in brewing and it might be converted to some unknown products which might also be toxic. The most important stages of beer production process having an inhibitory impact on AFB1 levels are malting, mashing or fermentation. During these stages, this contaminant is either removed with drainage water, spent grains and fermentation residue, diluted or destroyed as a result of thermic treatment, or adsorbed on the surface. In this way, detoxication methods such as heat and fermentation processes can go along way in reducing the levels of AFB1. In general, in order that barley germinate and grow (malt) satisfactorily, and therefore meet the requirements of the malting industry, it must be stored under conditions which would not promote the growth of storage molds. Nevertheless, the most important tools for the prevention and limiting of the appearance of AFB1 in foods are good practice in agricultural production and food storage. Foods should be stored at relative temperatures and humidities that discourage both fungal growth and AFs production.

It is important to carry out a strict quality control of AFB1, hence the need to detect this compound in foods and feedstuffs have motivated the development of analytical methods aiming to their quantification in beer, especially chromatographic methods, immunologic methods, sensors and biosensors, etc. The attempt of innovation in the development of new methods of analysis is dedicated towards the choice of new receptors (protein, nucleotides, MIPs), strategy of transduction (electrochemical, optical), miniaturization (lab-on-chip, microfluidic) or the use of novel materials such as nanoparticles. They may lead to lowering of LODs (up to the level of pg mL^{-1}) and broad linear ranges.

Strategies of AFB1 decontamination and prevention can be applied at all production stages: fungicide treatments on the field, lactic acid bacteria during malting and brewing, special yeast strains, ozonation, hot water treatment of barley grains etc. They are needed because sometimes commodities contamination is inevitable and the economic loss of it is too high to be discarded. However, we cannot forget that beers brewed from quality, purified and well stored raw materials do not represent any health risk. If raw materials comply with the limits fixed by the EC Regulation, the contribution of a moderate daily consumption of beer to AFB1 intake does not contribute significantly to the exposure of the consumer. Anyway, further studies are needed on the impact of beer production process on mycotoxin levels in order to better understand the risk to the population.

References

Álvarez-Segovia, K., García-Varela, R., Sergio García, H., Aguilar-Toala, J., Estrada-Montoya, M. C., Vallejo-Córdoba, B., González-Córdova, A. F. and Hernández-Mendoza, A. (2019). Determination of AFB1 in clear lager beer samples from Mexico and the possible correlation between physicochemical parameters and AFB1 levels. *Journal of Consumer Protection and Food Safety*, in press (https://doi.org/10.1007/s00003-019-01238-0).

Anastassiades, M., Lehotay, S. J., Stajnbaher, D. and Schenck, F. J. (2003). Fast and easy multiresidue method employing acetonitrile extraction/partitioning and "dispersive solid-phase extraction" for the determination of pesticide residues in produce. *Journal of AOAC International*, 86: 412-431.

AOAC Official Method 994.08 (2000). Aflatoxins in corn, almonds, Brazil nuts, peanuts, and pistachio nuts, multifunctional column (Mycosep) method. Natural toxins-chapter 49 (pp. 26-27). *Official Methods of*

Analysis of AOAC International, 17th edition, volume II, AOAC International, Gaithersburg, Maryland, USA.

Beloglazova, N. V. and Eremin, S. A. (2015). Rapid screening of aflatoxin B1 in beer by fluorescence polarization immunoassay. *Talanta*, 142: 170-175.

Benešová, K., Běláková, S., Mikulíková, R. and Svoboda, Z. (2012). Monitoring of selected aflatoxins in brewing materials and beer by liquid chromatography/mass spectrometry. *Food Control*, 25: 626-630.

Bertuzzi, T., Rastelli, S., Mulazzi, A. and Amedeo, G. D. (2011). Mycotoxin occurrence in beer produced in several European countries. *Food Control*, 22: 2059-2064.

Burdaspal, P. A. and Legarda, T. M. (2013). Survey on aflatoxin in beer sold in Spain and other European countries. *World Mycotoxin Journal*, 6: 93-101.

Chauhan, R., Singh, J., Sachdev, T., Basu, T. and Malhotra, B. D. (2016). Recent advances in mycotoxins detection. *Biosensensors & Bioelectronics*, 81: 532-545.

Chu, F. S., Chang, C. C., Ashoor, S. H. and Prentice, N. (1975). Stability of aflatoxin B1 and ochratoxin A in brewing. Applied Microbiology, **29:** 313-316.

Du, L. J., Chu, C., Warner, E., Wang, Q. Y., Hu, Y. H., Chai, K. J., Cao, J., Peng, L. Q., Chen, Y. B., Yang, J. and Zhang., Q. D. (2018). Rapid microwave-assisted dispersive micro-solid phase extraction of mycotoxins in food using zirconia nanoparticles. *Journal of Chromatography A*, 1561: 1-12.

Frenich, A. G., Vidal, M. J. L., Gonzalez, R. R. and Luiz, M. A. (2009). Simple and high-throughput method for the multi-mycotoxin analysis in cereals and related foods by ultra-high performance liquid chromatography/tandem mass spectrometry. *Food Chemistry*, 117: 705-712.

Ghali, R., Hmaissia-khlifa, K., Ghorbel, H., Maaroufi, K. and Hedili, A. (2008). Incidence of aflatoxins, ochratoxin A and zearalenone in Tunisian foods. *Food Control,* 19: 921-924.

González-Jartín, J. M., Alfonso, A., Rodríguez, I., Sainz, M. J., Vieytes, M. R. and Botana, L. M. (2019). A QuEChERS based extraction procedure coupled to UPLC-MS/MS detection for mycotoxins analysis in beer. *Food Chemistry*, 275: 703-710.

Goud, K. Y., Sharma, A., Hayat, A., Catanante, G., Gobi, K. V., Gurban, A. M. and Marty, J. L. (2016). Tetramethyl-6-carboxyrhodamine quenching-based aptasensing platform for aflatoxin B1: Analytical performance comparison of two aptamers. *Analytical Biochemistry*, 508: 19-24.

Juan, C., Berrada, H., Mañes, J. and Oueslati, S. (2017). Multi-mycotoxin determination in barley and derived products from Tunisia and estimation of their dietary intake. *Food and Chemical Toxicology*, 103: 148-156.

Khan, M. R., Alothman, Z. A., Ghfar, A. A. and Wabaidur, S. M. (2013). Analysis of aflatoxins in nonalcoholic beer using liquid-liquid extraction and ultraperformance LC-MS/MS. *Journal of Separation Science*, 36: 572-577.

Lulamba, T. E., Stafford, A. and Njobeh, P. B. (2019). A sub-Saharan African perspective on mycotoxins in beer – a review. *Journal of the Institute of Brewing*, 125: 184-199.

Lv, X., Li, Y., Cao, W., Yan, T., Li, Y. and Du, B. (2014). A label-free electrochemiluminescence immunosensor based on silver nanoparticle hybridized mesoporous carbon for the detection of aflatoxin B1. *Sensors & Actuators B*, 202: 53-59.

Mably, M., Mankotia, M., Cavlovic, P., Tam, J., Wong, L., Pantazopoulos, P., Calway, P. and Scott, P. M. (2005). Survey of aflatoxins in beer sold in Canada. *Food Additives and Contaminants*, 22: 1252-1257.

Matumba, L., Poucke, C., Biswick, T., Monjerezi, M., Mwatseteza, J. and Saeger, S. (2014). A limited survey of mycotoxins in traditional maize based opaque beers in Malawi. *Food Control*, 36: 253-256.

Molina-García, L., Fernández-de Córdova, M. L. and Ruiz-Medina, A. (2012). Indirect determination of aflatoxin B1 in beer via a multi-commuted optical sensor. *Food Additives and Contaminants*, 29: 392-402.

Nigussie, A., Bekele, T., Gemede, H. F. and Woldegiorgis, A. Z. (2018). Level of aflatoxins in industrially brewed local and imported beers collected from Ethiopia market. *Cogent Food & Agriculture*, 4: 1453317.

Odhav, B. and Naicker, V. (2002). Mycotoxins in South African traditionally brewed beers. *Food Additives & Contaminants*, 19: 55-61.

Oliveira, P., Brosnan, B., Jacob, F., Furey, A., Coffey, A., Zannini, E. and Arendt, E. K. (2015). Lactic acid bacteria bioprotection applied to the malting process. Part II: Substrate impact and mycotoxin reduction. *Food Control*, 51: 444-452.

Oliver G. (2011). The Oxford Companion to Beer, 1st edition. *Oxford University Press.*

Pagkali, V., Petrou, P. S., Makarona, E., Peters, J., Haasnoot, W., Jobst, G., Moser, I., Gajos, K., Budkowski, A., Economou, A., Misiakos, K., Raptis, I. and Kakabakos, S. E. (2018). Simultaneous determination of aflatoxin B1, fumonisin B1 and deoxynivalenol in beer samples with a label-free monolithically integrated optoelectronic biosensor. *Journal of Hazardous Materials*, 359: 445-453.

Pascari, X., Ramos, A. J., Marín, S. and Sanchís, V. (2018). Mycotoxins and beer. Impact of beer production process on mycotoxin contamination. A review. *Food Research International*, 103: 121-129.

Piacentini, K. C., Savi, G. D., Pereira, M. E. V. and Scussel, V. M. (2015). Fungi and the natural occurrence of deoxynivalenol and fumonisins in malting barley. *Food Chemistry*, 187: 204-209.

Pietri, A., Bertuzzi, T., Agosti, B. and Donadini, G. (2010). Transfer of aflatoxin B1 and fumonisin B1 from naturally contaminated raw materials to beer during an industrial brewing process. *Food Additives and Contaminants*, 27: 1431-1439.

Peters, J., van Dam, R., van Doorn, R., Katerere, D., Berthiller, F., Haasnoot, W. and Nielen, M. W. F. (2017). Mycotoxin profiling of 1000 beer samples with a special focus on craft beer. *PLoS One*, 12: 0185887.

Romero-González, R., Martínez-Vidal, J. L., Aguilera-Luiz, M. M. and Garrido-Frenich, A. (2009). Application of conventional solid-phase extraction for multimycotoxin analysis in beers by ultrahigh-performance liquid chromatography–tandem mass spectrometry. *Journal of Agricultural and Food Chemistry*, 57: 9385-9392.

Rudrabhatla, M., George, J. E., Hill, N. R. and Siantar, D. P. (2008). Detection of relevant mycotoxins in wheat beer and sake by LC-MS/MS using prototype immunoaffinity column clean-up: a preliminary study. *Food Contaminants: Mycotoxins Food Allergens*, 1001: 241-254.

Scott, P. M. and Lawrence, G. A. (1997). Determination of aflatoxins in beer. *Journal of AOAC International*, 80: 1229-1234.

Shetty, P. H., Hald, B. and Jespersen, L. (2007). Surface binding of aflatoxin B1 by *Saccharomyces cerevisiae* strains with potential decontaminating abilities in indigenous fermented foods. *International Journal of Food Microbiology*, 113: 41-46.

Sun, L., Wu, L. and Zhao, Q. (2017) Aptamer based surface plasmon resonance sensor for aflatoxin B1. *Microchimica Acta*, 184: 2605-2610.

Tan, Y., Chu, X., Shen, G. L. and Yu, R. Q. (2009). A signal-amplified electrochemical immunosensor for aflatoxin B1 determination in rice. *Analytical Biochemistry*, 387: 82-86.

Tamura, M., Uyama, A. and Mochizuki, N. (2011). Development of a multi-mycotoxin analysis in beer-based drinks by a modified QuEChERS method and ultra-high-performance liquid chromatography coupled with tandem mass spectrometry. *Analytical Science*, 27: 629-635.

Turner, N. V., Subrahmanyam, S. and Piletsky, S. A. (2009). Analytical methods for determination of mycotoxins: a review. *Analytica Chimica Acta*, 632: 168-180.

Ventura, M., Guillén, D., Anaya, I., Broto-Puig, F., Lliberia, J. P., Agut, M. and Comellas, L. (2006). Ultra-performance liquid chromatography/ tandem mass spectrometry for the simultaneous analysis of aflatoxins B1, B2, G1, G2 and ochratoxin A in beer. *Rapid Communications in Mass Spectrometry*, 20: 3199-3204.

Visconti, A., Pascale, M. and Centonze, G. (2000). Determination of ochratoxin A in domestic and imported beers in Italy by immunoaffinity clean-up and liquid chromatography. *Journal of Chromatography A*, 888: 321-326.

Wei, S., Liu, Y., Yan, Z. and Liu, L. (2015). Molecularly imprinted solid phase extraction coupled to high performance liquid chromatography for determination of aflatoxin M1 and B1 in foods and feeds. *RSC Advances*, 5: 20951.

Xu, G., Zhang, S., Zhang, Q., Gong, L., Dai, H. and Lin, Y. (2016). Magnetic functionalized electrospun nanofibers for magnetically controlled ultrasensitive label-free electrochemiluminescent immune detection of aflatoxin B1. *Sensors and Actuators B*, 222: 707-713.

Zachariasova, M., Cajka, T., Godula, M., Malachova, A., Veprikova, Z. and Hajslova, J. (2010). Analysis of multiple mycotoxins in beer employing (ultra)-high-resolution mass spectrometry. *Rapid Communications in Mass Spectrometry*, 24: 3357-3367.

Zhao, Y., Huang, J., Ma, L., Liu, S. and Wang, F. (2017). Aflatoxin B1 and sterigmatocystin survey in beer sold in China. *Journal of Food Additives & Contaminants: Part B*, 10: 64-68.

Zollner, P. and Mayer-Helm, B. (2006). Trace mycotoxin analysis in complex biological and food matrices by liquid chromatography-atmospheric pressure ionisation mass spectrometry. *Journal of Chromatography A*, 1136: 123-169.

INDEX

A

acetic acid, 51, 85, 88, 91, 93, 133
acetonitrile, 88, 90, 91, 139, 140, 145, 151
acid, 5, 16, 19, 49, 51, 67, 88, 89, 90, 132, 148, 154
acidity, 19
active compound, 17, 21
adaptation, 8
adsorption, 5, 23, 135
adverse effects, x, xiii, 87, 125, 128
AFB1, xiii, 126, 128, 129, 134, 135, 136, 137, 138, 141, 144, 145, 146, 147, 148, 149, 150, 151
aflatoxin, viii, ix, x, xiii, 25, 33, 72, 74, 75, 87, 117, 120, 125, 126, 151, 152, 153, 154, 155, 156, 157
aflatoxin B1, viii, ix, x, xiii, 1, 33, 72, 74, 75, 87, 117, 120, 125, 126, 151, 152, 153, 154, 155, 156, 157
Africa, 122
agar, xii, 45, 46, 58, 59, 60, 84, 85, 91, 104, 106, 118, 119
aggregation, 17
aging process, 134
agriculture, 87, 144
air temperature, xi, 40
algorithm, 47
alkaloids, 127
almonds, 151
amine, 140
ammonium, 51, 85
amphibians, 22
amylase, 131, 132
analytical methods, 55, 70, 126, 141, 150, 156
annealing, 93
antibiotic, 91
Argentina, 27
arthropods, 22
Asian countries, 135
assessment, 26, 68, 72
atmosphere, 13, 60
Austria, 24
authority, 28, 114
automation, 141

B

B1 (AFB1), xiii, 125, 128
bacteria, 22, 26, 132, 148, 150, 154
bacterial cells, 18

bacterial strains, 16, 18
barriers, xi, 40, 43
base, 93
beer, viii, x, xiii, 18, 125, 126, 128, 129, 130, 131, 132, 133, 134, 135, 136, 137, 138, 140, 141, 142, 143, 144, 145, 148, 149, 150, 151, 152, 153, 154, 155, 156, 157
Belgium, 49, 51, 80
beneficial effect, xi, 40
benefits, 117
beverages, 128, 129, 146
bicarbonate, 50, 51
bile, 18
biocompatibility, 20
biodegradability, 20
biopreservation, 15, 22, 24, 31
biosensors, 147, 150
biosynthesis, 35
biotechnology, 17, 36
biotic, 86
blood, 3, 11, 64
bone, 117
bone marrow, 117
Brazil, 120, 151
breast milk, 5, 29, 35
breeding, 6
brevis, 16
brewing, x, xiii, 126, 128, 129, 130, 131, 132, 134, 135, 136, 141, 142, 148, 149, 150, 152, 153, 155
Bulgaria, 121
by-products, 21

C

Ca^{2+}, 20
cacao, 130
calibration, 49, 52, 54, 56, 139, 148
cancer, 128
capillary, 52, 141
carbon, 133, 140, 153
carbon dioxide, 133
carcinogen, x, xiii, 86, 125, 128
carcinogenicity, 86, 127
CBS, 47, 75, 76
CCA, xii, 84, 85, 91, 103, 106, 114
CH3COOH, 50, 51
cheese, x, 2, 7, 8, 9, 21, 25, 27, 29, 30, 72
cheese and meat, x, 2, 21
chemical, 13, 14, 21, 23, 26, 79, 80, 81, 132, 147, 148
chemical reactions, 132
chemicals, 49, 50
chicken, 30, 71
China, 5, 157
chitin, 20
chitosan, 20, 24, 27
chlorine, 8
chloroform, 88, 89, 90, 92, 93, 106
chromatograms, 55, 61, 62, 63, 95, 96, 98
chromatography, xii, 84, 141
chromosome, 117
chronic kidney disease, 118
citrinin, vii, x, xii, 31, 67, 72, 83, 84, 85, 87, 107, 117, 118, 119, 120, 121, 122, 135
civilization, 129
clarity, 134
classes, 126, 144
classification, 86
cleanup, 47, 139
climate, 66, 112, 114
climates, 127
CO2, 60, 133, 134
cocoa, 12, 86, 127
coffee, 3, 6, 12, 86, 130
color, v, 89, 131, 132
combined effect, 117
commercial, 10, 17, 32, 135
commodity, 15, 86, 134
community, 11
competition, 15, 17, 127
complexity, 138, 146

composition, 5, 10, 14, 23, 127
compounds, 14, 15, 16, 17, 20, 126, 130, 133, 137, 138, 139, 140, 146
conditioning, 134
Congress, v, 74, 80
consumers, 6, 9, 13, 66
consumption, xii, 11, 26, 35, 41, 44, 76, 84, 87, 113, 116, 129, 136, 137, 140, 151
contaminant, 10, 149
contaminated food, 41, 114
contamination, ix, xi, xiii, 3, 4, 5, 6, 7, 9, 10, 11, 13, 19, 23, 28, 32, 33, 34, 40, 44, 56, 64, 65, 67, 68, 69, 73, 74, 75, 86, 87, 91, 95, 102, 112, 113, 114, 116, 117, 119, 121, 122, 126, 127, 128, 129, 132, 148, 149, 150, 154
control policy, 2
controversial, 19
cooking, 13, 131
cooling, 130, 133, 136
copolymers, 139
correlation, 151
cost, 130, 139, 145, 146, 147
cotton, 46, 109
cracks, 42
critical value, 127
Croatia, ix, 39, 44, 45, 79, 80, 82
crop, 3, 22, 126
crops, 86, 87
crust, 42
CTAB, 85, 92
CTO, 143
culture, 19, 23, 25, 92, 103
culture medium, 23
cycles, 47, 93
cycling, 47

D

dairy products, ix, x, 2, 3, 4, 16, 18, 25, 37
damages, v, 66
database, 47, 72, 76, 94, 108
decontamination, ix, x, 2, 4, 11, 12, 14, 15, 17, 23, 68, 126, 148, 149, 150
degradation, 6, 15, 16, 17, 87
Delta, 88
denaturation, 18, 93, 147
Denmark, 64
depth, 8, 9
detection, x, xi, xiii, 5, 6, 9, 28, 29, 30, 40, 44, 49, 50, 52, 66, 75, 119, 121, 126, 137, 138, 141, 142, 143, 144, 145, 146, 147, 152, 153, 157
detection techniques, x, xiii, 126, 144
detoxification, 5, 6, 30, 36, 128, 148
developing countries, 5
diet, xi, 27, 40, 69, 134
dietary intake, 153
diffusion, 43, 66
diseases, 87, 128
distilled water, 48, 91
distribution, 3, 7, 86, 87, 103, 114, 126
DNA, 2, 46, 92, 93, 94, 108, 117, 119
DNA polymerase, 94
DOI, 29, 31, 37
drainage, 149
draught, 134
drying, ix, x, 2, 3, 10, 13, 19, 21, 41, 42, 45, 66, 75, 131

E

E. coli, 17
economic losses, 126
Egypt, 5, 9, 30, 37, 118
election, 139
electrophoresis, 47, 93, 141
ELISA, xi, 5, 40, 48, 49, 50, 53, 55, 65, 137, 145
ELISA method, 49, 50, 53, 55, 65
elongation, 93
elucidation, 116

emission, 91, 106, 145, 146
energy, 19, 52
engineering, 17
enlargement, 18
environment, xi, 3, 5, 19, 40, 43, 59, 127
environmental conditions, 10, 34, 138
environments, 18, 42, 59, 65, 67
enzymatic activity, 148
enzyme, 17, 131
enzymes, 15, 16, 36, 42, 129, 131, 132
equipment, 144, 148
ESI, 52
ethanol, 93, 133
ethyl acetate, 48, 88, 89
ethylene, 119
ethylene oxide, 119
Europe, 122, 135
European Commission, x, 2, 70, 80, 113, 118
European Union, 70, 76, 82, 113, 118, 127
evidence, 7, 9
evolution, 133
excitation, 90, 145, 146
exclusion, 21
excretion, 8
execution, 130
exposure, 6, 7, 26, 68, 72, 76, 114, 115, 136, 137, 151
extraction, 46, 49, 50, 95, 119, 120, 132, 140, 142, 143, 144, 145, 147, 151, 152, 153, 155, 156
extracts, 90, 94, 95, 96, 98, 99, 100, 101, 102, 106, 107
exudate, 46

F

false positive, 145
farmers, 149
farms, 6
FAS, 113
fat, 72
feedstuffs, 118, 150
fermentation, xiii, 13, 16, 42, 126, 130, 133, 135, 136, 148, 149
filters, xi, 40, 43
filtration, 17, 51, 89, 135, 136
financial, 116
fitness, 8
flavonoids, 34
flavor, 10, 23, 73, 129, 131, 132
flavour, 42, 56, 76, 130, 132
flight, 144
flora, 6, 7
flour, 66, 87
fluorescence, 89, 91, 104, 144, 146, 147, 152
food, ix, x, xi, xii, xiii, 2, 3, 4, 6, 11, 12, 13, 14, 15, 18, 19, 20, 22, 23, 26, 27, 28, 30, 31, 32, 34, 36, 40, 41, 44, 45, 52, 64, 69, 71, 75, 76, 78, 79, 80, 81, 84, 86, 87, 103, 112, 113, 114, 116, 118, 119, 122, 126, 127, 128, 134, 139, 148, 149, 150, 152, 157
food additive(s), xiii, 20, 126, 134
food chain, 28
food industry, 13
food products, xi, 2, 4, 113, 116, 128
food quality, 2, 14, 79, 80, 81
food safety, ix, x, 2, 14, 27, 28, 29, 79, 80, 81, 122
food spoilage, 2
force, 80, 81
formation, 2, 18, 19, 21, 131
formula, 35
France, 72
fungi, xii, 13, 14, 15, 21, 22, 23, 25, 28, 29, 32, 35, 75, 85, 86, 87, 88, 91, 102, 107, 108, 109, 111, 112, 114, 118, 120, 126, 135
fungus, 13, 23, 24, 127

G

gamma radiation, 30
gel, 47, 50, 51, 89, 93, 140, 146
genes, 34, 70, 76, 111, 122
genetics, 122
genre, 140
genus, 41, 42, 43, 46, 56, 57, 102, 117
Germany, 5, 11, 45, 46, 47, 48, 53
germination, 20, 131
global scale, 44
glucose, 15
glycerol, 133
glycoproteins, 23
graduate students, 78
graphite, 140
GRAS, 20
grass, 5
growth, 7, 8, 14, 15, 19, 20, 21, 22, 23, 25, 33, 34, 35, 37, 42, 46, 56, 59, 66, 68, 75, 112, 114, 127, 128, 133, 148, 150
guidelines, x, 2

H

habitats, 8
harvesting, 112
hazards, 25, 26, 68, 79, 80, 81
haze, 133
health, xiii, 11, 41, 66, 67, 85, 86, 114, 115, 116, 122, 151
health effects, 86, 114
hepatotoxicity, xii, 84, 87, 127
hexane, 88, 90, 145
High Pressure Liquid Chromatography (HPLC), xii, 30, 49, 50, 84, 85, 88, 89, 90, 92, 95, 99, 100, 101, 102, 106, 107, 114, 142, 143, 144
history, 122, 128
homeostasis, 19
human, x, xii, xiii, 11, 21, 29, 33, 35, 41, 43, 70, 84, 86, 116, 121, 125, 129, 134, 137, 149
human body, 21, 41
human exposure, 70, 121
human health, 43, 121, 129, 149
humidity, xi, 10, 14, 40, 43, 45, 127
hydrogen, 16, 19
hydrogen bonds, 19
hydrogen peroxide, 16
hydrolysis, 132
hydroxyl, 19
hydroxyl groups, 19

I

identification, xiii, 34, 36, 45, 46, 47, 52, 68, 85, 102, 108, 115, 144
illumination, 47
immersion, 46
immunosuppression, 127
impairments, 41
in vitro, 12, 17, 23, 36, 118, 147
in vivo, 86
incidence, 4, 13, 30, 71, 114
incubation period, 19
India, ix, xii, 84, 87, 88, 94, 112, 113, 115, 117, 118, 119, 120
industry, 22, 29, 49, 52, 150
infants, 6
ingestion, 6, 10, 134
ingredients, xi, 19, 40, 129
inhibition, 14, 16, 17, 19, 22, 23, 27, 35
injury, v
inoculation, 46, 69
integrity, 15
interference, 46
ion exchangers, 139
ions, 52, 55
Ireland, 74
irradiation, 14, 24, 28

isolation, 59, 91, 102
issues, ix, x, 2, 79, 128
Italy, 1, 7, 9, 11, 32, 64, 71, 72, 73, 156

J

Japan, 90

K

kidney, 115, 116
kidneys, 72

L

lactic acid, 148, 150
Lactobacillus, 25, 27
LC-MS, xi, 40, 49, 50, 52, 54, 55, 142, 143, 153, 155
LC-MS/MS, xi, 40, 49, 50, 52, 54, 55, 142, 143, 153, 155
lead, 18, 19, 148, 150
Lebanon, 7, 28
legislation, 64, 113, 141
light, 22, 131
lipases, 16
lipolysis, 76
liquid chromatography, xi, 40, 141, 144, 152, 155, 156, 157
liquid phase, 136
liver, 11, 116
livestock, 87
loci, 47
longevity, xi, 40, 43
LTD, 50, 51
Luo, 31
lysis, 92

M

magnesium, 140
Maillard reaction, 18
majority, 129
maltose, 132
management, 122
manufacturing, 10, 24, 43, 69, 126
market share, 130
Maryland, 151
mass, xi, 40, 51, 52, 55, 148, 152, 155, 156, 157
mass spectrometry, xi, 40, 152, 155, 156, 157
materials, 136, 139, 140, 141, 142, 151, 152
matrix, 7, 8, 56, 129, 138, 139, 140, 145, 146
matter, v
measurement(s), 52, 56, 76, 146
meat, ix, x, xi, xii, 2, 3, 8, 10, 13, 16, 18, 19, 20, 21, 22, 23, 24, 25, 28, 29, 30, 31, 32, 33, 34, 35, 36, 40, 41, 42, 43, 44, 45, 48, 49, 50, 52, 53, 54, 56, 58, 61, 64, 65, 66, 67, 68, 70, 71, 72, 73, 74, 75, 76, 79, 80, 81
meat products, ix, xi, 2, 3, 8, 10, 13, 16, 19, 20, 21, 23, 24, 25, 28, 29, 32, 33, 35, 36, 40, 41, 42, 43, 44, 45, 48, 50, 53, 54, 56, 58, 61, 64, 65, 66, 67, 68, 70, 71, 72, 73, 74, 75, 76, 79, 81
media, 7, 37, 46
median, 9, 137
medical, 116
Mediterranean, 44
Mediterranean countries, 44
metabolism, 36, 71
metabolites, 14, 21, 43, 86, 128
methanol, 50, 51, 88, 139, 146
methodology, 140
Mexico, 135, 151
mice, 117

microorganisms, 7, 12, 14, 15, 17, 19, 21, 22, 148
microscope, 46
microscopy, 46
microwaves, 12
migration, 26
milk, x, 2, 3, 4, 5, 6, 16, 25, 26, 29, 30, 31, 35, 37
miniaturization, 145, 150
mixing, 48, 89, 93
model system, 35
modifications, 48
moisture, 3, 12, 42, 60, 128
moisture content, 3, 12
mold, ix, xii, 2, 4, 9, 15, 17, 20, 23, 71, 84, 89, 91
molds, 3, 4, 7, 10, 14, 17, 19, 21, 22, 24, 73, 126, 150
molecular weight, 138
molecules, 23, 147
monomers, 139
morphology, 91, 110
mortality, 127
mortality rate, 128
moulds, xi, 24, 25, 30, 40, 41, 42, 43, 45, 56, 64, 65, 66, 67, 68, 69, 71, 74, 75
muscles, 73
mycelium, 46
mycotoxin-producing fungi, 13, 85, 91, 114, 126, 135
mycotoxins, vii, x, xii, xiii, 3, 7, 10, 11, 12, 13, 14, 15, 16, 21, 23, 25, 26, 27, 28, 29, 31, 32, 33, 35, 37, 41, 42, 43, 64, 65, 66, 67, 68, 69, 70, 71, 72, 74, 75, 83, 84, 85, 86, 87, 96, 112, 113, 115, 116, 118, 119, 120, 121, 125, 126, 127, 128, 129, 134, 136, 137, 140, 144, 148, 149, 152, 153, 154, 155, 156, 157
Mycotoxins, vii, 10, 15, 25, 28, 29, 32, 33, 68, 71, 72, 83, 85, 86, 126, 149, 154, 155

N

NaCl, 3, 14, 23, 34, 93
nanofibers, 157
nanomaterials, 140
nanoparticles, 140, 147, 150, 152
National Research Council, 1
nebulizer, 52
negative effects, 42
nephropathy, xii, 84, 87, 115, 121, 122
Netherlands, 47
nitrogen, 50, 51, 145
Norway, 5
novel materials, 150
nucleic acid, 47, 147
nucleotides, 150
nutrients, 14, 20
nutrition, 29, 81

O

ochratoxin A, 1, vii, ix, x, 1, 2, 25, 26, 27, 28, 29, 30, 31, 32, 33, 34, 35, 36, 37, 39, 41, 48, 50, 54, 58, 59, 60, 68, 69, 70, 72, 73, 83, 85, 117, 119, 121, 122
Ochratoxin A (OTA), vii, ix, x, xi, xii, 1, 2, 3, 4, 5, 6, 7, 8, 9, 10, 12, 13, 14, 15, 16, 18, 19, 20, 22, 23, 25, 26, 27, 28, 29, 30, 31, 32, 33, 34, 35, 36, 37, 39, 40, 41, 43, 44, 45, 48, 49, 50, 51, 52, 53, 54, 55, 56, 57, 58, 59, 60, 61, 62, 63, 64, 65, 66, 67, 68, 69, 70, 71, 72, 73, 74, 75, 76, 83, 84, 85, 86, 87, 88, 89, 90, 91, 92, 94, 95, 96, 97, 98, 99, 100, 101, 102, 104, 106, 107, 109, 111, 112, 113, 114, 115, 116, 117, 118, 119, 120, 121, 122, 127, 129, 135, 136, 137, 152, 156
ochratoxin family, 2
oil, 46, 87, 113
optimization, 55
organism, 8, 110

overlap, 146
oxidation, 132
oxidative stress, 36
oxygen, 13, 16
ozonation, 148, 150
ozone, 12

P

pancreas, 16
participants, 80
pathway, 15, 23
PCR, xi, 40, 47, 70, 76, 85, 93, 94, 108, 111, 119, 122
peptide, 25, 147
peptides, 22
permission, v
peroxide, 16
pesticide, 140, 151
pH, 18, 19, 42, 51, 72, 92, 93, 127, 132, 148
phenolic compounds, 19
phosphate, 50, 51
photodegradation, 146
photographs, 106
phylogenetic tree, 94, 110
physical treatments, 12, 14
pigs, 27, 33, 65, 69, 73
plants, 10, 22, 24, 36, 69
platform, 153
polar, 140
polarization, 152
policy, 2
polymerase, xi, 40
polymerase chain reaction, xi, 40
polymerization, 139
polymers, 139
polyphosphates, 81
population, 7, 23, 72, 87, 115, 151
Portugal, 6, 70, 135
potassium, 88, 90
potato, 91
poultry, 80, 81, 122, 128
precipitation, 134
preparation, v, 17, 48, 55, 71, 81, 93, 129, 139
preservation, 14
prevention, vii, ix, x, 1, 2, 4, 12, 32, 43, 65, 71, 127, 128, 149, 150
principles, 43, 145
probiotic, 36
probiotics, 18
processing stages, 13
producers, xii, 11, 27, 40, 43, 56, 57, 59, 66, 112, 115, 121, 149
production technology, 45
profitability, 37
project, 68
proliferation, 138
proteinase, 93
proteins, 15, 20, 21, 132, 133, 134
proteolytic enzyme, 22
protons, 20
prototype, 155
PSA, 140
public health, xiii, 41, 126
purification, 90, 95

Q

quality assurance, 80
quality control, 79, 150
quality improvement, 80
quantification, 25, 49, 53, 112, 139, 145, 150
quantum dot, 147

R

radiation, 12, 14, 30
raw materials, 33, 45, 73, 128, 135, 136, 149, 151, 155
reactions, 145

reactive oxygen, 16
reagents, 48, 88
real time, 91
receptors, 150
recommendations, v
recovery, 49, 53, 54
regions of the world, 9, 11
regression, 56
regulations, 81, 113, 118, 127, 128, 136
regulatory bodies, 43, 64
rehydration, 10
remediation, ix, x, 2, 3, 32
remediation strategies, ix, x, 2, 3, 32
repression, 15, 28
reproduction, 7
requirement(s), 52, 55, 138, 140
residue, 149
residues, 13, 151
resolution, 144, 157
respiration, 19
retail, 119
ribosomal RNA, 76, 122
rights, v
risk, xiii, 13, 24, 34, 37, 43, 85, 112, 114, 115, 116, 123, 128, 137, 149, 151
risk assessment, 123
risks, 112, 119
room temperature, 93, 134
routes, 3, 4, 10

S

safety, x, xiii, 2, 23, 28, 29, 42, 44, 80, 81, 82, 126
salts, 18, 140
science, 73, 77, 79, 80, 81
SDS, 92
seed, 113
selectivity, 139, 140, 144, 146
sensitivity, 22, 56, 140, 141, 144, 145
sensor, 146, 154, 156
sensors, 146, 150
sequencing, 47, 76, 94, 122
services, v
shape, 46, 111, 139
sheep, 5, 29, 30
showing, 56, 95, 96, 98, 99, 100, 101, 102, 149
side chain, 19
significance level, 53
silica, 89, 140, 146
silver, 153
SLE, 143, 144
smoking, 19, 45, 60
sodium, 35, 50, 51, 93, 140
software, 49, 91, 94, 111
solid phase, 152, 156
solubility, 136
solution, 27, 48, 49, 50, 51, 90, 91, 146
solvents, 49, 88, 139, 147
sorption, 140
South Africa, 135, 154
South America, 128, 135
Spain, 5, 11, 125, 135, 137, 152
species, x, 2, 4, 6, 7, 8, 9, 16, 17, 18, 21, 22, 23, 24, 35, 41, 46, 47, 56, 57, 59, 65, 70, 86, 103, 112, 117, 118, 121, 127, 139
stability, x, xiii, 27, 126, 147
stabilization, 138
standard deviation, 49
standardization, 71
starch, 87, 129, 131
state(s), 64,81
sterile, 45, 91, 93
stock, 49
stomach, 5
storage, ix, x, 2, 3, 4, 10, 13, 33, 41, 64, 65, 67, 86, 87, 116, 126, 127, 128, 148, 150
structure, 46
sub-Saharan Africa, 138, 153
substrate(s), 15, 17, 21, 34, 48, 86, 112, 115
sucrose, xii, 85, 92
sulfate, 140

sulfur, 133
Sun, 26, 70, 148, 156
suppliers, 145
supply chain, 32
surface area, 46
Sweden, 5
Switzerland, 47, 116
synthesis, 112, 115, 147

T

target, 136
TDI, 7, 113
teams, 81
techniques, 15, 31, 78, 79, 129, 134, 148
technologies, 37, 44, 72, 79, 144
temperature, 3, 10, 12, 19, 23, 25, 42, 43, 51, 52, 72, 74, 87, 114, 126, 127, 131, 132
texture, 46
TGA, 93
thermal stability, 147
thinning, 18
tobacco, 130
toluene, 88, 89
toxic effect, 127
toxicity, xii, 20, 26, 36, 84, 87
toxicology, 72, 128
toxin, 3, 7, 8, 12, 15, 17, 20, 25, 96, 104, 135, 136, 137, 138, 147, 149
trademarks, 137
traditional dry-cured meat products, xii, 40, 41, 73
transcriptomics, 35
transduction, 145, 150
transfer, 65, 126, 129, 134, 135, 155
transmission, 64
transportation, 4
treatment, x, xiii, 12, 15, 18, 65, 126, 135, 138, 139, 148, 149, 150
trifluoroacetic acid, 145
tuition, 78, 79
tumours, 121
Turkey, 5, 37

U

United States, 26
urinary tract, 121
USA, 7, 47, 48, 51, 88, 93, 94, 151
UV light, 89
UV-irradiation, 146

V

vacuum, 51, 89, 90, 92
validation, 53, 54, 145
valve, 146
varieties, 133
vegetable oil, 127
Vietnam, 113, 120
vinyl chloride, 119
vitamin E, 119

W

waste, 8, 149
water, 25, 42, 47, 48, 50, 51, 60, 72, 74, 89, 90, 92, 93, 127, 129, 131, 132, 140, 145, 146, 149, 150
wavelengths, 145, 146
web, 77
wells, 48
WHO, 71, 87, 113
Wisconsin, 47
World Health Organization, 71, 122, 149
World Health Organization (WHO), 122, 149
worldwide, xi, xiii, 40, 86, 87, 112, 126, 149

Y

yeast, xii, 10, 14, 18, 19, 23, 32, 46, 85, 92, 129, 130, 133, 134, 149, 150

Z

zirconia, 140, 152